Leaving the field

Leaving the field

Methodological insights from ethnographic exits

Edited by Robin James Smith and
Sara Delamont

MANCHESTER UNIVERSITY PRESS

Copyright © Manchester University Press 2023

While copyright in the volume as a whole is vested in Manchester University Press, copyright in individual chapters belongs to their respective authors, and no chapter may be reproduced wholly or in part without the express permission in writing of both author and publisher.

Published by Manchester University Press
Oxford Road, Manchester M13 9PL

www.manchesteruniversitypress.co.uk

British Library Cataloguing-in-Publication Data
A catalogue record for this book is available from the British Library

ISBN 978 1 5261 5765 2   hardback

First published 2023

The publisher has no responsibility for the persistence or accuracy of URLs for any external or third-party internet websites referred to in this book, and does not guarantee that any content on such websites is, or will remain, accurate or appropriate.

Typeset
by New Best-set Typesetters Ltd

Contents

List of contributors	page vii
Leaving the field: an editors' introduction *Sara Delamont and Robin James Smith*	1

Part I Entanglements and im/perfect exits

1 Finishing fieldwork in less than perfect circumstances: lessons learned in 'labyrinth' exiting *Alexandra Allan and Sarah Cole*	33
2 Exeunt omnes!! The case for bad exits in ethnography *Sally Campbell Galman*	45
3 Reflections on care and attachment in the 'departure lounge' of ethnography *Alex McInch and Harry C.R. Bowles*	62
4 Unfinished business: a reflection on leaving the field *Gareth M. Thomas*	74
5 Materia erotica: making-love among glassblowers *Erin O'Connor*	86

Part II Troubling the field

6 Those who never leave us *Jessica Nina Lester and Allison Daniel Anders*	101
7 *Déjà vu et jamais vu*: what happens when the field expands in ways that mean there is no exit? *Dawn Mannay*	113
8 Student voices 'echo' from the ethnographic field *Janean Robinson, Barry Down and John Smyth*	126

9 Public space and visible poverty: research fields without exit 139
   Andrew P. Carlin
10 'The martial will never leave your bones': embodying
   the field of the Kung Fu family 152
   George Jennings

Part III Intermissions and returns

11 Between open and closed: recursive exits and returns to
   the fuzzy field of a community library across a decade
   of austerity 167
   Alice Corble
12 On the importance of intermissions in ethnographic
   fieldwork: lessons from leaving New York 180
   Joe Williams
13 Can you remember? Leaving and returning to the field
   in longitudinal research with people living with dementia 192
   Andrew Clark and Sarah Campbell
14 A constant apprenticeship in martial arts: the messy
   longitudinal dynamics of never leaving the field 205
   David Calvey

Part IV Returns, responsibilities and representations after 'leaving'

15 A cautionary tale about 'respondent validation': the dissonant
   meeting of 'field self' and 'author self' 219
   Daniel Burrows
16 Commenting on legal practice: research relationships and
   the impact of criticism 231
   Daniel Newman
17 Emotional honesty and reflections on problematic positionalities
   when conducting research in another country 243
   Ashley Rogers

Index 255

List of contributors

Robin James Smith is Reader in Sociology at Cardiff University. His research is concerned with talk, embodied action, and categorisation practices. He has studied interaction in public space, traffic order, outreach with the street homeless, and the organisation of mountain rescue work. His most recent project is a study of the use of visual technology in police accountability, oversight, and training. He is the editor of *On Sacks* (Routledge, 2021), *The Lost Ethnographies* (Emerald, 2019), and *Urban Rhythms* (SAGE, 2013).

Sara Delamont is Reader Emerita in Sociology at Cardiff University. She currently conducts fieldwork on *capoeira*, the Brazilian dance-fight game, and *savate*, the French kick-boxing martial art. Her previous books include *The Lost Ethnographies* (Emerald, 2019), *Embodying Brazil: An Ethnography of Diasporic Capoeira* (Routledge, 2017), *Key Themes in the Ethnography of Education* (SAGE, 2014), *Feminist Sociology* (SAGE, 2003), and *Fieldwork in Education Settings* (3rd ed., Routledge, 2016). Together with Paul Atkinson, she was the founding editor of the journal *Qualitative Research* (SAGE) and the SAGE *Research Methods Foundations* resource. She was one of the editors of the *SAGE Handbook of Ethnography* (SAGE, 2001). She is a Fellow of the Academy of Social Sciences and a Fellow of the Learned Society of Wales. She is the recipient of the Lifetime Achievement Awards from the British Sociological Association and the British Research Education.

Alexandra Allan is currently the Head of the School of Education at the University of Exeter. Her research interests reside in the field of the Sociology of Education. Alexandra's research has primarily focused on issues relating to educational inequalities. She has a particular interest in gender and academic achievement, sexualities, and social class. This research has been published in a number of journals, including *Gender and Education, Discourse: Cultural Studies in the Politics of Education, Feminist Theory* and *Qualitative Research*. Much of this work will be drawn together in a forthcoming book, due to be published with Routledge in 2023.

Allison Daniel Anders, PhD, is an associate professor in the educational foundations and inquiry programme and the qualitative research certificate programme at the University of South Carolina. She teaches foundations of education, critical race theory, social theory, and introductory and advanced qualitative research methodologies. Dr Anders studies critical and postcritical qualitative research methodologies, contexts of education, and the everyday experiences of targeted youth. Her research includes work with children and families with refugee status, LGBTQ+ students and educators, and youth who experience incarceration.

Harry Bowles is a Lecturer in the Department for Health at the University of Bath and Fellow of the Higher Education Academy. He is a member of the Centre for Equality in Sport, Physical Activity and Health, contributing to research that aims to increase understanding of the individual, socio-cultural and structural factors that influence opportunities and experiences in sport and physical activity across the life course. Harry's research focuses on the sociology of youth and physical culture and has examined topics related to gambling, physical education, and young people's transitions into employment in and through sport.

Dan Burrows is a former social worker and now senior lecturer in social work at Cardiff University. He undertook ethnographic research with a hospital social work team for his professional doctorate, which formed the basis for his first book, *Critical Hospital Social Work*, which was published in 2020. His teaching and research interests include unpaid carers, social work with older people and rights-based practice.

David Calvey is a senior lecturer in sociology at Manchester Metropolitan University (MMU). Prior to working at MMU he held teaching, visiting, and research positions at the University of Manchester, Liverpool John Moores University, the Open University, and the University of Queensland, Australia. His publications and expertise range across covert research, situated ethics, humour studies, sensory ethnography, martial arts, deviance, ethnomethodology, bouncers, and organisational creativity. He is a member of the British Sociological Association (BSA) and a Fellow of the Higher Education Academy. He won the Emerald Literati award for outstanding paper in 2022. His latest book is *Covert Research: The Art, Politics and Ethics of Undercover Fieldwork* (Sage, 2017).

Sarah Campbell is a senior lecturer at Manchester Metropolitan University within Integrated Health and Social Care. Her research interests are in ageing, dementia, gender, long-term care, sensory and embodied experiences, and social inequalities. She works with participatory qualitative and creative methods and ethnography. Her most recent work includes Uncertain Futures, a participatory arts-led research study exploring inequalities around both paid and unpaid work for older women in Manchester.

Andrew P. Carlin, PhD, is a research consultant who teaches bibliometrics and information work at Ulster University. He has published in various peer-reviewed journals, focusing particularly on ethnomethodology, information, and textual work. His research interests include interdisciplinarity, the praxeology of information, and the problem of data as a phenomenon of order. His interests in public space and studies of urban environments take into account such diverse topics as libraries and information agencies, liminal spaces, security and terror, and street ethnography.

Andrew Clark is a Professor in the School of Health and Society at the University of Salford. His research explores everyday experiences of neighbourhoods and communities of place using innovative and participatory approaches. He is currently investigating the development of age-friendly neighbourhoods since the Covid-19 pandemic, how people living with dementia experience local places, and the development of training and support for the care-home workforce.

Sarah Cole is currently a lecturer in education at the School of Education at the University of Exeter. She teaches and researches the broad area of social justice in education with a specific interest in gender and gender-based violence in education. This work has been published in the edited collection *International Perspectives on Exclusionary Pressures in Education* (Palgrave Macmillan, 2023) and in the *Journal of Gender-Based Violence*.

Alice Corble is an interdisciplinary library and archival scholar, educator, librarian, and activist. Her scholarship and practice explores the hidden infrastructures and power dynamics of library spaces, collections, and intersectional social relations. She is currently based at the University of Sussex Library where she teaches critical information literacy alongside an AHRC-RLUK fellowship researching the integral role of the library and archives in understanding the university's institutional origins, development, and contemporary calls to decolonise.

Barry Down is Adjunct Professor in the Centre for Research in Educational and Social Inclusion (CRESI) at the University of South Australia and Emeritus Professor, Murdoch University. He has been Chief Investigator on a number of Australian Research Council (ARC) grants investigating student engagement, school-to-work transitions, early career teacher resilience and performance arts. His current research involves a critical policy analysis of school exclusion policies.

Sally Campbell Galman

Sally Campbell Galman is Professor of Child and Family Studies at the University of Massachusetts @ Amherst. She is a visual artist and arts-based researcher and the author/illustrator behind the *Shane the Lone Ethnographer* comic research methods texts and lots of other fun stuff. Keep up with her projects at www.sallycampbellgalman.com ♥

George Jennings is Senior Lecturer in Sport Sociology at Cardiff Metropolitan University, where he leads the MA Sport, Ethics in Society course. An avid practitioner, George has been researching the martial arts through ethnography and other qualitative research designs since his undergraduate dissertation (2004–05), including studies of Wing Chun, Taijiquan, Xilam and historical European martial arts (HEMA). George is founder of the Wales Martial Arts Practitioner–Researcher Network and co-convenor of the Documents Research Network (DRN). He is currently learning the Chinese-American art of Cheng Hsin through a private apprenticeship.

Jessica Nina Lester, PhD, is a Professor of Qualitative Methodology in the School of Education at Indiana University, Bloomington. At Indiana University, she serves as the Program Coordinator of the Qualitative and Quantitative Methodology PhD program and oversees the Certificate in Qualitative Research and Inquiry Methodology. Jessica is a qualitative methodologist and interdisciplinary researcher who has published extensively in the field of qualitative inquiry and more specifically ethnomethodology and conversation analysis (EMCA). In her methodological scholarship, she focuses on the study and development of EMCA, the integration of digital tools and spaces in qualitative research, and the place of disability in critical qualitative inquiry. In much of her substantive research, she has sought to examine clinical and educational interactions that involve children and youth. At Indiana University, she teaches qualitative methods courses and mentors graduate students in qualitative inquiry from a range of disciplines.

Dawn Mannay is a Reader (Associate Professor) in the School of Social Sciences at Cardiff University, with interests in education, inequalities and children and young people. Dawn's recent books include *Children and young people 'looked after'? Education, intervention and the everyday culture of care in Wales* (University of Wales Press, 2019) co-edited with Louise Roberts and Alyson Rees, *The Sage Handbook of Visual Research Methods* (SAGE, 2020) co-edited with Luc Pauwels, and *Creative Research Methods in Education* (Policy Press, 2021) co-authored with Helen Kara, Narelle Lemon and Megan McPherson. Dawn is committed to working creatively with communities to produce data and disseminate the messages from research findings in innovative ways to increase the potential for social, educational and policy change and support informed practice.

Alex McInch is the Professional Doctorate Coordinator in the Cardiff School of Sport and Health Sciences at Cardiff Metropolitan University in Wales, UK. He is a social policy researcher with a specific interest in social inequality. His work focusses predominantly on the sport and education policy spaces and alongside research activity, he is a keen Doctoral supervisor, examiner, and chair of examination boards. Alex's current projects have been commissioned by several local authorities in Wales and have looked at inequalities in both health and education.

Daniel Newman is Reader in law at Cardiff University with research expertise on access to justice. His books include *Legal Aid Lawyers and the Quest for Justice* (Hart, 2013), *Justice in a Time of Austerity* (Bristol University Press, 2021) with Jon Robins, *Experiences of Criminal Justice* (Bristol University Press, 2022) with Roxanna Dehaghani, and *Legal Aid and the Future of Access to Justice* (Hart, 2023) with Jacqueline Kinghan, Jess Mant and Catrina Denvir. He edited *Leading Works in Law and Social Justice* (Routledge, 2021) and *Access to Justice in Rural Communities* (Hart, 2023) with Faith Gordon, and *Leading Works on the Legal Profession* (Routledge, 2023).

Erin O'Connor is an Associate Professor of Sociology and Chair in the Department of Politics and Human Rights, as well as Affiliated Faculty of Environmental Studies in the Department of Natural Sciences. Her research specialises in work and labour, art and craft, knowledge and culture, the body and environment and is guided by material feminism, critical indigenous theory, phenomenology, critical ecological theory, and post-humanism. Her book manuscript, *Firework: Art, self, and world among glassblowers*, draws from four years of ethnographic research in a glassblowing studio to analyse contemporary craft in industrial and knowledge economies. Dr O'Connor is a 2023 recipient of the Rakow Grant for Glass Research at the Corning Museum of Glass. She has published in multiple leading journals and international monographs. Dr O'Connor enjoys the outdoors, creating, and sharing her interests with her children (while learning a lot about superheroes and Legos along the way).

Janean Robinson is an Honorary Research Associate at the Centre for Research in Educational and Social Inclusion (CRESI) at the University of South Australia as well as Murdoch University. A former high school teacher for 30 years, Janean's research focuses on the changing nature of teachers' work and student behaviour management regimes in neoliberal times. She has co-authored *Rethinking school-to-work transitions in Australia: Young people have something to say* and is currently working on a new book, *The 'trouble' with school behaviour discipline policies in neoliberal times.*

Ashley Rogers is a Lecturer in Criminology at the University of Stirling. She currently conducts research on issues surrounding rights, forced migration, indigenous knowledges and extreme weather events in Scotland, Malawi, and Zimbabwe. She is also an Associate Director of the Scottish Graduate School of Social Sciences supporting postgraduate student engagement. Ashley's teaching focuses on critical criminology and in particular, crimes of the powerful. She delivers methods training on doing ethnography and is rather nostalgic about the gift of time offered by the PhD to conduct such immersive research. She completed her PhD in 2017 and immediately entered a lectureship at Abertay University where she remained until 2022.

John Smyth is Emeritus Research Professor, Federation University Australia. He is a Fellow of the Academy of the Social Sciences in Australia, a former Senior Fulbright Research Scholar, and the recipient of several awards from the American Educational Research Association for his critical ethnographic work. He has been a university academic for 51 years and is the author/editor of over 40 books and more than 300 scholarly papers. He is a sociologist who has worked on a variety of aspects of critical sociology and social justice. His most recent book is *The Toxic University: Zombie Leadership, Academic Rock Stars, and Neoliberal Ideology* (Palgrave Macmillan, 2017). John is also the founder and current series editor of the Palgrave Critical University Studies series.

Gareth Thomas is a Senior Lecturer in the School of Social Sciences at Cardiff University. He is a sociologist interested in disability, medicine, health/illness, reproduction, and stigma. His latest project – funded by the British Academy as part of its Mid-Career Fellowship scheme – explores how learning-disabled people craft alternative and affirmative accounts of their lives, which depart from popular (and problematic) narratives of deficit, tragedy, and dependence.

Joe Williams is an early career researcher at Cardiff University. He recently completed a PhD in sociology, an ethnography of homeless outreach teams in Manhattan, spending a year as part of an outreach team. His research is concerned with how understandings of urban homelessness are established and operationalised via the 'doing of' outreach work, and the implications this has for street-based care work, and for clients/service-users traversing systems of provision. He is currently a research associate at Y-lab, Cardiff University.

Leaving the field: an editors' introduction

Sara Delamont and Robin James Smith

Introduction

The literature concerning the fieldworker's position within the field and their positionality in relation to their informants has proliferated. The traditional 'don't ask, don't tell' approach to the production of field observations has given way to a far more reflexive mode of operating; a more sociable, less extractive and more roundly ethical engagement with the lifeworlds of those engaged with through ethnographic research (Sinha and Back 2014). Indeed, almost every stage of fieldwork has been used as the basis for autobiographical accounts of projects. These are sometimes called 'confessionals', which we regard as a somewhat pejorative term following Atkinson's (1996; 2017) analysis and do not use here. As this genre has grown, nearly all the stages of fieldwork and many other aspects of ethnography have been explored. Examples abound of such autobiographical, revelatory and sometimes comical accounts of early days in the field: establishing rapport with informants, ethical dilemmas, data collection, data analysis and writing. Yet, despite the growth of autobiographical accounts of fieldwork, two common ethnographic experiences remain relatively neglected. We have taken it upon ourselves to gather writings that attend to these overlooked areas in two separate but related collections.

The first of these overlooked experiences concerns what we called 'lost' projects; insights from fieldwork that had ended abruptly or had never even properly begun or had not been written up had been largely ignored. In 2017, when we commenced that project, we had been unable to discover any coherent collection of papers on failures or analyses of the causes of the loss and of the methodological consequences of ethnographies that never took place. And so we edited a collection examining stories of 'lost' projects in order to explore the methodological insights that could be gained from recovering and re-examining fieldwork 'disasters' (Smith and Delamont 2019). By examining apparent 'failures' in field research, we hoped not only to provide encouragement to novice researchers, who might be experiencing difficulties with their own research, that even failures can provide material for publication, but to trouble some of the assumptions as to what a 'good' piece of field research might look like. The second overlooked experience, which is, for obvious reasons, more common than research 'failing' is *leaving the field*. This aspect of fieldwork, at once central and peripheral, and the lessons that can be learned from ways in which ethnographers have conducted, or have been forced to negotiate, their *ethnographic exits* is the focus of this collection.

The contrasting experiences of leaving the field have been emphasised by Amanda Coffey (1999: 55), who wrote:

> While for some the ending of fieldwork is a welcome respite, for others the leaving of the place, and more often the people, can cause sadness, anguish and pain.

Coffey (1999) draws on the accounts of Cannon (1992) and Blackwood (1995) to illustrate her statement. Cannon had studied women with breast cancer and felt she could not ethically and emotionally simply terminate her relationships with them. Blackwood had been in Indonesia and, as well as the usual feelings of loss, had a very personal exit narrative. She had fallen in love with a woman and could not in that era bring her lover legally to the US because lesbian relationships were treated differently from heterosexual ones.

Good researchers think about their exits, just as they think about their first days. Unlike accessing the field, however, the 'culture' of ethnography has not mandated routinely writing about leaving, or the thinking associated with it, in publications. Many authors do not discuss *how* they left their field sites in their published work at all. Most textbooks say little or nothing about doing ethnographic exits, let alone what might be learned from them. Indeed, Coffey (1999: 106) criticises textbook authors Lofland and Lofland (1995) for reducing the exit to a procedural matter, noting that 'From this advice it would appear that ending fieldwork is a necessity to be skilfully managed, rather than an experience to be lived through.' Coffey (1999: 106) goes on to rehearse several other aspects of exiting, because: 'The ending of fieldwork and leaving the field usually represents the end of a particular phase of an ethnographer's life and career', and 'fieldwork becomes, and is part of, who we are and will always be'. (Coffey 2018: 74)

The very act of leaving, we argue, enables ethnographers to recall and reflect upon the places, the people, the time and the particular and often peculiar sense of self encountered during fieldwork. Leaving and *returning* can also mean getting to know the field in a new way in addition to reading through, revisiting and analysing our fieldnotes and transcripts, as well as in our writing practices. While we might recognise in that process the analytic insights yielded by leaving the field, readers are frequently denied access to the moments that generate those insights. Some exceptions have been published (discussed below), but we remain convinced that there is more ground to cover in terms of how many aspects of the whole ethnographic experience can be illuminated by a closer attention to leaving the field. Put another way, we are convinced that ethnographic exits provide intellectually valuable data allowing for meaningful inquiry into often taken-for-granted aspects of the ethnographic enterprise, including 'field relations', 'research ethics' and 'the field' itself. Following some initial discussion with colleagues (who all responded positively), this collection was planned and authors were recruited. There was a deliberate decision not to include chapters about our own field exits, although we do reflect below on our own experiences of (not quite) leaving. It is also worth noting that the collection was planned and in progress before COVID-19 swept across the world and, as such, there are no chapters on its disruptive effects on fieldwork, despite this being a globally repeated phenomenon (although see Smith et al. 2020 for a reflection on the impact of COVID-19 on fieldwork and research ethics). The remainder of this introduction is in four parts: we establish the neglect of exits in the ethnographic literature and review what literature there is; we argue our case for analysing exits as rigorously as every other phase of fieldwork; then we draw out what insights we can from some of our own exits; and finally, we very briefly outline the structure and contents of the book.

Exit stage, left?

The literature on the topic of exits is sparse, and what does exist is little cited, so the highlights are briefly summarised before our self-explorations. We do not rehearse what textbooks such as Delamont (2016a), Coffey (1999) or Lofland and Lofland (1995) advise novices to do but, instead, endorse Coffey's (1999: 105) verdict that such books offer practical advice rather than reflection. We aim for the latter.

Coffey (1999; 2018) and Delamont (2016a) both emphasise the relative lack of material on exits, based on their impressions of the literature; but a systematic review in a paper based in, and written for, scholars in management, business and organisational studies by Michailova, Piekkari, Plakoyiannaki, Ritvala, Mihailova and Salmi (2014) provides confirmation of the topic's scarcity. These authors conducted a systematic review of the literature from the 1970s to 2010. Their review covered (a) encyclopaedias of research methods; (b) ten social science textbooks written for management and organisation researchers; (c) special issues on research methods in seven core journals in management and organisation published between 1979 and 2010; (d) all the issues of six journals focused on qualitative methods (such as *Qualitative Market Research*) from 1995 to 2010; (e) the 'editorial notes' on writing up qualitative research from the *Academy of Management Journal*; and (f) 'statements of ethical practices'. This last category is not explained, but presumably the authors examined the ethical codes produced by learned societies such as the British Psychological Society.

Michailova and co-authors conclude that: 'knowledge of what constitutes exit from fieldwork is missing' (p 142). When it is mentioned, the coverage is 'simplistic' and it is treated 'largely a physical process'. They judge that:

> The underlying view is that exiting is a rational, instrumental process, and the emphasis is on how exit should be 'executed' and 'managed' ... Such an approach, besides being overly simplistic, focuses solely on the researcher – it is geared toward her/his self-interest and short-term personal gains. (pp 142–143)

Their conclusions favour Coffey's emphasis on thinking more broadly, and deeply, about leaving the field. We regard the conclusions drawn from this systematic review as vindicating our decision to curate this collection. Our thoughts on the key literature now follow, mentioning a few insightful contributions, and we return to Michailova et al. at the end of this section.

An early attention to leaving the field was offered by Snow (1980), who also argued for more *analytic* attention to be paid to exits. Few scholars took Snow's advice, until Iversen (2009) focused on leaving the field. Her account of 'Getting out' of an ethnographic project on poverty is a landmark study. Following that article, Iversen (2019) wrote the entry for the Sage *Research Foundations* encyclopaedia. That entry is a clearly written overview of the sparse literature available, and we do not recapitulate it here. Rather, we focus on chapters in two largely forgotten collections from 1980 and 1991 focusing on aspects of the literature which explore issues also raised in the chapters in this collection or offer useful advice for inexperienced ethnographers so that they can think about their own exits. The topic was a designated section comprising four chapters in Shaffir, Stebbins and Turowetz's (1980) collection *Fieldwork Experience*, and there is a parallel section with five chapters in Shaffir and Stebbins' *Experiencing Fieldwork* (1991). Few collections on ethnographic methods contain such sections, perhaps for good reason. Ethnographic research monographs which show every sign of thoughtful and self-critical investigations by their authors are often silent about the end of the work. Bosse (2015), Davis (2015) and Vannini (2012), for example do not describe respectively the terminations of their studies of ballroom dancing in Illinois, tango in Argentina and The

Netherlands or the everyday lives of people who live on islands off the west coast of Canada and depend on the ferries for their education, shopping, travel to work and social lives. Boellstorff (2008) says little about leaving the online field site of Second Life. A handful of ethnographers have, however, explored the 'reverse' culture shock of their homecomings: Barley (1983), for example, writes humorously of his return from fieldwork in West Africa and, more recently, Alice Goffman (2014) described something of her return to 'white middle-class society' as revealing aspects her embodied experience of being in the field on 6th Street (although the account itself is not without its representational troubles ...).

The Shaffir and Stebbins (1991) collection contains five papers on leaving and keeping in touch by Wolf (1991), Gallmeier (1991), Kaplan (1991), Taylor (1991) and Stebbins himself (1991) which complement the four chapters in Shaffir, Stebbins and Turowetz (1980). In that earlier collection Roadburg (1980), Letkemann (1980) and Altheide (1980) have individual papers and Maines, Shaffir, Haas and Turowetz (1980) report on their own experiences and summarise a correspondence about exits which they conducted with senior ethnographic experts of that era: Rosalie Wax, Prudence Rains, Anselm Strauss, Donald Roy, Howard Becker and Herbert Gans. The ideas of the leading scholars of 1946–79 might now seem rather old fashioned, and some of the experts may even have been forgotten, but the problems they faced are not irrelevant today. That is why some of their work is revisited in Atkinson, Coffey and Delamont (2003). Rosalie Wax told Maines and his colleagues the story of two very different exits: one happy and centred on a ceremony, the other a clandestine escape at night because of a government edict issued to the organiser of the programme Wax was supposed to be researching that demanded her expulsion. There are some parallels here with the two contrasting exits discussed by Allan and Cole in Chapter 1. Altheide's (1980) essay is particularly useful for a novice, because it focuses on productive use of the end days of the research period. He argues that the fieldwork period immediately prior to leaving can be very fruitful. He had been studying television newsrooms in California and argues that once his exit date was known, he was able to capitalise on that certainty to get data that had previously been hard to obtain. Deadlines can, indeed, focus the mind. Some informants agreed to be interviewed in his last weeks, and he was able collect material by stressing it was his last chance to obtain it.

Other endings are occasioned by the 'field' leaving the ethnographer. Maines et al. (1980) reports an 'exit' of that type. He had been doing a study of postdoctoral fellows in an American university and he writes insightfully about how, precisely because the 'post doc' is a transitory status, his informants routinely moved on to new jobs in other cities leaving *him* 'behind'. In other words, his 'field' left him. A similar phenomenon, the 'field' being populated by actors who leave, and the effects that has on the ethnographer, is reported by Ilane Kaplan (1991) who did fieldwork among fishermen, as did Carolyn Ellis (1986). Kaplan points out that 'Comings and goings fit the life-style of fishermen who are regularly out at sea' (Kaplan 1991: 234), and so her departures were unremarkable to her informants. Robin (Smith) imagines that when he, eventually, leaves the mountain rescue team, his departure will be treated similarly; people leave for all sort of reasons, and he'll be just one more.

The chapters in Shaffir and Stebbins (1991) are slighter than those in Shaffir, Stebbins and Turowetz (1980), but address exits from field sites that exemplify the variety of settings in which ethnographers work. Two of them explore male sporting activities and have some parallels with the Chapters 10 and 14 in this collection. Gallmeier (1991) and Stebbins (1991) both include fieldwork on male professional sportsmen: Stebbins on a Canadian football team and Gallmeier an ice hockey team. In both cases the rhythm of the sports season provided a 'natural' end for the fieldwork. In Chapters 10 and 14 of this collection their authors both reflect on individual physical activities (martial arts) rather than team sports, and as amateur enthusiasts rather than full-time professionals, but the camaraderie and the performative masculinity observed have similarities.

Gallmeier (1991) spent the 1981–82 American ice hockey season embedded in a minor league men's team. American professional ice hockey has four tiers, with the National Hockey League, the Majors, at the top and three 'minor' levels below. Gallmeier's team, 'The Rockets' were in the third tier. He had his own nickname (The Scarecrow) and spent many hours at rinks, in changing rooms, on the tour bus and hanging out in bars. The fieldwork ended 'naturally' when the ice hockey season did, and the men all dispersed. Gallmeier got a job in another part of the US but was able to keep up with the team because his family lived in The Rockets' home town and he could combine family duty and reunions with the players he had studied. He reports that as the season drew to a close he experienced feelings of 'alienation, guilt and sadness' (Gallmeier 1991: 226). However, he was also 'looking forward to disengaging from the Rockets' because he felt 'a cultural clash' (p. 227) between himself and the sexism, violence, practical jokes and homosocial atmosphere of the team. Gallmeier argues that leaving, revisiting and staying in touch, which he did by subscribing to *Hockey News*, all, in different ways, have advantages for improving the original data, but only if those data are analysed using the exit, any revisits and other contacts in academic ways.

Wolf (1991), who studied a biker gang in Canada, writes more about his access than his exit but he addresses the issue of fieldwork which is all-encompassing when the researcher is in it, but utterly out of reach when the investigator leaves. He writes usefully about how, because the bikers were very self-contained, tightly organised and socially marginal, he was able to do the research only by being totally immersed in their world. Wolf (1991: 222) thought that after he had 'finished' his main data collection he 'would at least maintain ties of friendship' and that 'the enduring emotion would be one of comradeship'. He realised, however, that the engagement had to be all-embracing or entirely non-existent, and so when he left he quickly became an ex-member 'who had simply drifted away', as other men did. In a neat phrase, Wolf remarks how he was 'quickly reclaimed by everyday life'.

Several authors in the two collections advise novices on how to know when to leave. Steven Taylor (1991) focuses on an ethnographic study he conducted in a secure hospital on a ward warehousing seventy-three young men with major learning disabilities, and a long-term study of a couple surviving life in extreme poverty. Chapter 4 in the present collection draws explicitly on this. Taylor (1991: 242) explores the 'how do you know when to stop?' question in an informative way. His first suggestion is that a project is nearing its end 'when one can begin to recognise the puzzle and the pieces fit together' and 'the data become repetitious'. More personally, he confesses that he knows the end of a study is near 'when I become bored writing fieldnotes' (p. 243). However, he advises that when that stage is reached, 'Firstly, stay a while longer', because 'research suffers when it is rushed or concluded prematurely' (p. 243). Taylor has learned that once he is becoming bored with writing fieldnotes it is time to check carefully that the fieldwork has been thorough. The ethnographer should 'think what you may have missed before you conclude your study' (p. 243). In his study in the total institution, he was, in the late stages of the project, able to focus on an aide who had been reluctant to talk to him thus far, and finally convinced the man to be interviewed. Taylor (1991) also addresses a commonly reported, but under-analysed, feature of ethnographic research. He is the only one of the five authors in Shaffir and Stebbins' (1991) collection to point out to the inexperienced ethnographer that informants may well have forgotten that the ethnographer is a researcher, and feel abandoned and even angry when the investigator leaves. This is a point made by Hammersley and Atkinson (2019) and Coffey (1999), and is reflected on, below, in Robin's tale of leaving.

Gallmeier's (1991) guilt and sadness at leaving The Rockets is contrastive with the experiences of Stebbins (1991). He reflects on twelve projects he has done on jazz, classical music, magicians, stand-up comics, Canadian football (which is more like American football than any other variety), astronomy, archaeology, baseball and the theatre, eight of which were based on participant observation. Stebbins has himself performed as a magician, a stand-up comedian and a musician. He says that when he packs up and goes home after fieldwork, he is generally very happy, because while he starts projects with enthusiasm, he is used to them ending with fatigue and theoretical saturation. Stebbins's pleasure in leaving the field is partly because he finds that writing and publication is 'the most interesting phase' (p. 248) of his investigations. His contribution to the Shaffir and Stebbins collection says little else about the processes of exit but focuses on varieties of engagement after the fieldwork. He has, for example, been involved in negotiations to lower Canadian customs duty on imported telescopes (Gallmeier 1991: 253) as a contribution to the ongoing lives of amateur astronomers he had met initially as a researcher. He does, however, stress that some researchers 'may reach the end of the data collection phase of their research unaware that they have done so' (p. 249).

Fine (1983;1985; 1987; 1996; 1998; 2013; 2018) has, like Stebbins, done ethnographies in many settings, and therefore has left many field sites; from Little League baseball and vocational training for catering students to competitive chess players and graduate studies in fine art. Fine's (1983) account of deciding it was time to leave the field, in his monograph about the Golden Brigade fantasy game club – a central part of a wider project in the then novel field of fantasy gaming – suggests a parallel thought process to Stebbins'. Fine discovered that, because the high level of turnover at the Golden Brigade fantasy game club (so called because it was in a room over a fire station), he had become one of the 'experts', expected to explain rules to novices and run games rather than just play them, and that he was not learning anything new. This combination of having become too familiar (this is not a phrase Fine himself uses) and too much of a 'full participant' meant that, for Fine, the sociological 'pay off' had declined sharply. While we recognise Fine's decision as being the correct one in the context of his broader project, we also question what 'too familiar' might mean in ethnographic studies concerned with knowledge of a skilled or esoteric practice (see Smith 2022) which, in turn, raises questions of how to 'leave' a field that has been written on to the body through practice (see Chapters 10 and 14).

A small body of literature has also considered *returns* to the field. Burawoy (2003), for example, extols the virtues of focused revisits to document processes of social change and to ' focus on the inescapable dilemmas of participating in the world we study, on the necessity of bringing theory to the field, all with a view to developing explanations of historical change' (p. 647). Comparison with the findings of previous scholars is argued to be an important part of developing rigorous theoretical ethnography. Anne Oakley (2016) reports on a return to re-interview participants in her original

'Becoming a Mother' study. Interestingly, given that the original publications from the project emphasised friendship and deep trust between interviewer and interviewees, a significant proportion of the women interviewed did not recall the project or even, in one case, Oakley's being present at the birth of their child. It is a reminder, among other things, that ethnographers should not assume too great a significance for their fleeting presence in the lives of others.

There are other sorts of returns and relationships to data and 'the field' that we also think worth studying but which are, unfortunately, absent from this collection. The sorts of possibilities and contingencies raised by cross-generational encounters in field returns are one example. The relationship between different researchers producing and re-using secondary data is another. Nigel Fielding (2004) produces an interesting discussion of the relationship between primary fieldwork and secondary data analysis in terms of the analytic and practical management of context effects. We tried, but failed, to recruit a conversation analyst to the present collection to write about the strangely intimate but distanced field relation that endures from initial video data collection, on through the hundreds of repeated viewings by the researcher and others in data sessions, and even across academic generations in the context of 'famous' data recordings that are often returned to and reanalysed.

We now return to that systematic review by Michailova et al. (2014). The authors use the findings of their systematic review as the basis for an argument about the relationships between types of exits and their potential for formal theorising. They provide a two-by-two table, creating four possible relationships between the exit and the potential for theory. They use one axis to separate high from low potential for theorising, and the other to divide exits which produce disruptive versus sustained research relationships. It is not made clear why researchers would want, or choose, sustained research relationships (see our personal stories below!), and there seems no category in their model for a transitory field site. Table 1 presents our version of their figure 2 (Michailova et al. 2014: 150).

The authors propose that the Anticipated Exit produces conventional theorising, the Hostage Exit one-sided theorising, the Revelatory Exit leads

Table 1 Types of exit and their potential for theory

Impact on research	Major	Minor
Sustained relationships	1 Anticipated exit	2 Hostage exit
Disrupted	3 Revelatory exit	4 Black hole exit

to paradoxical theorising and the Black Hole exit to 'blind alley theorising' – that is, often no theory at all. Their example of such an exit is Belousov, Horlick-Jones, Bloor, Gilinskiy, Golbert, Kostikovsky, Levi and Pentsov (2007). In that project a key research gatekeeper was murdered in Russia and the project had to be abandoned. We are not convinced by the label 'black hole', because the common misconception is that black holes suck in and destroy everything. Nothing is left and nothing can exist. We see this as a *very* unlikely outcome from an ethnographic project. And, as it goes, black holes do not actually 'suck' anything into them and *do* spew out a good deal of material. It seems that the trouble is *recognising* that material spewed from the event horizon in the first place, and then recognising that it might be useful for ethnographic insight and theory. So, while we are impressed by the findings of the systematic review, we do not share the conclusions of Michailova et al. As the chapters in this collection demonstrate, there is something to be said of all sorts of exits (and, indeed, as we have demonstrated previously, even of exits from projects that 'never happened'). Dramatic, flawed, awkward, painful, abrupt and unexpected exits can *all* provide insight, data and opportunities for theorising and conceptual development. Hence, our argument put forward in this introduction for exiting as an *active* and *attentive* ethnographer.

Exiting as an *active* ethnographer

One reason for this collection is precisely to prepare readers such that they do not simply drift out of their fieldwork without self-critical, and *analytic*, reflections on what they are doing. Before we explain our case for all types of fieldwork to include a self-conscious and analytic attention to exiting, a caveat applies to all the published accounts, and to our own which follow in the next section. The critical and analytic attitude should be extended to the reading of authors' published autobiographies and autobiographical accounts of their fieldwork. Such accounts should never be read naively, as if they are literally 'true'. They are a genre with its own conventions, as analysed by Atkinson (1996; 2017) and Van Maanen (2011). Such accounts often focus on the exciting, the dramatic, and frequently on things that went wrong. Dick Hobbs (1988), for example, stresses how he wore the wrong clothes and drank too much, Lincoln Keiser (1969) recounts how he nearly got shot in a gunfight between gangs and, (in)famously, Alice Goffman (2014) admits to driving a car with an informant as passenger, toward what could be seen as an attempted murder. Given these conventions for producing the account of the research, 'the literature' on exits is likely to give a novice a negative and highly coloured overview of terminating fieldwork, rather than a balanced one. Because being thrown out of a field site is dramatic there are probably more accounts of that in the confessional literature than there are reflections on planned and scheduled exits. And no doubt there are more accounts of dramatic exits in proportion to orderly ones in the literature than accurately reflect the real ratio of 'failed' fieldwork and messy exits to successful projects.

In short, we recommend reading about exits, but reading with the nature of the genre in mind. This attitude might be taken to the remainder of the book. While such tales might not be read 'literally', it is possible to gain methodological and empirical insights from 'failed' or 'lost' research projects: those that fail may have involved an unplanned or even enforced exit. The small literature on 'lost' projects includes stories of exits from the field, just as the small literature on leaving the field includes a few accounts of 'lost' projects. Two recent accounts of fieldwork that had to be terminated due to forces outside the ethnographers' control are Scott (2019), who reports on how a study of a swimming pool became impossible when it was privatised, and Campbell Galman (2019), who writes about how she had to leave a field site in India because she got too ill to work (along with a catalogue of other unfortunate events). Scott's piece is a good example of how economic and political events and forces far away from the ethnographer can destroy field sites, Campbell Galman's a timely reminder that the ethnographer's *body* is an essential factor in much fieldwork. The ethnographer's body figures centrally in several chapters in this collection, and in Chapter 5 in particular.

When fieldwork is brought suddenly to an end, through no fault of the ethnographer, there is no possibility of a planned and carefully managed 'leaving'. Scott's pool was closed to her: she did not decide that her research was at an end. But for many projects, the ethnographer does the 'leaving'. For physically distant sites, that is likely to be a physical as well as a social departure, as in Ashley Rogers' (Chapter 17) departure from Bolivia. For field sites that are close at hand, it can be more gradual, or a more episodic series of absences. As we have said, this is not the end of the research. Obviously, the thesis, the monograph, the papers all must be written. But, as time in the field goes on, we do need to think about our comfort zones. There are interpersonal comfort zones and intellectual ones, though often they coincide. It can be easy to get absorbed into a routine of fieldwork, observing and participating in the same round of activities, spending one's time with the same people, comfortable in the familiar places with the familiar people. Routine fieldwork of that sort means not having to renegotiate day-to-day access and social relations. It means in many cases that there is a ready-made framework for the fieldwork itself, such as an organisational schedule. At the same time, one can become thoroughly accustomed to what is being observed. School classrooms, court proceedings, clinics or opera rehearsals can all become thoroughly familiar. But it is easy to go on attending the site(s), the event(s), the gathering(s), accumulating 'data', but not acquiring any novel insights. That is what we mean by intellectual comfort zones.

Therefore, in advance of 'leaving' it is a good idea to review things. The main questions include: Have I got too comfortable, too 'at home'? Are there people, settings, events or things that I have neglected? Am I confining my field relations to a subset of actors I feel comfortable with, or have grown close to? Are there phenomena that I am now taking for granted, and need to revisit? The end of fieldwork – which of course is not the end of the project – is an important period for critical reflection.

There is a methodological rhetoric concerned with the 'completeness' of ethnographic fieldwork. It depends on analysing the data in the process of that fieldwork and relates to the idea of 'saturation', especially Glaser and Strauss's (1967) idea of theoretical sampling. Saturation is not about having a lot of data but is about the thorough exploration of one's key analytic ideas. As we construct analytic categories, we need to explore their dimensions, the extent of their application, to search for variations and deviant cases. We might add that when it comes to identifiable phenomena, it also means being able to document how they are accomplished, how they are identified by the actors and how – if applicable – they are evaluated by those in the field. 'Saturation' is, if anything, an ideal. Realistically, we recognise that most fieldwork involves compromises, and the perfect coverage of actors and events is virtually impossible. In the real world, we usually must make the most of what we can get, and elaborate plans about theoretical sampling are rarely matched by the reality. But there is a strong argument for reviewing the basics of our fieldwork before we leave the field and attempting to remedy manifest shortcomings. That means asking if we have done justice to different categories of actor, exploring the patterns and rhythms of everyday life and how they relate to our own fieldwork, or whether we have paid adequate attention to key phenomena, rather than treating them as obvious or 'given'.

In the next section we focus on some of our own exits, recognising that the end of fieldwork – which of course is not the end of the project – is an important period for critical reflection.

The editors' efforts at leaving the field

It is somewhat ironic that we are editing this compilation, because we are both very bad at leaving research sites and locations; perhaps that is a good enough motivation for gathering up the experiences of others! Here we tell our own tales of leaving the field.

Sara's story

I never intended to leave my two field sites in 2020. When COVID-19 put the UK into lockdown for many months, indeed I had planned to continue my fieldwork in them indefinitely, and had published a chapter on why that was so (Delamont 2016b). In that paper I had argued that actively doing fieldwork was a fundamental part of my sense of self, a point made in Delamont (2012). The impact of the coronavirus meant that the ethnographic research I was doing on two martial arts, capoeira and Savate, as they are taught and learned in the UK ended in ways unlike all the previous studies I had done since my PhD fieldwork in 1969–71. I had been studying classes and festivals of the African-Brazilian dance-fight-game capoeira since 2003, and of Savate (French kick boxing) since 2009. The capoeira research, conducted with Neil Stephens (Delamont, Stephens and Campos 2017) was done in several UK cities and focused particularly on three Brazilian masters (*mestres*) who are friends, as well as key informants. *Mestre* Claudio Campos, whom I had watched teach 681 two-hour lessons, is also a co-author of the monograph. I had so many friends among the teachers and students in UK capoeira that I intended to do a small amount of fieldwork until perhaps 2025 or 2026. The Savate project had begun as an exercise in contrastive theoretical sampling but turned into a project in its own right (Southwood and Delamont 2018a; 2018b; 2018c) and I had no plans to stop watching that at least until the Paris Olympics in 2024, when it may be a demonstration sport.

I retired from my full-time academic post in December 2014, but I continued to do fieldwork for two to six hours a week for about forty to forty-six weeks of the year after retirement, both to see friends and to keep active as an ethnographer (see Delamont 2016b). For so long as I am teaching methods, examining PhD theses, refereeing journal articles and writing about ethnography (Delamont and Atkinson 2021), I am sure that I *ought* to continue being an active researcher.

I observed a capoeira class in London on 5 March 2020, unaware that it would be my last for at least sixteen months. The teachers of both capoeira and Savate whom I know best very quickly took their classes online. These instructors earned their main living from their classes and three of them had dependent children, so they moved fast to keep their beloved martial art alive and active, not only for their own economic livelihoods but also for their own fitness, their own sanity, their students and their potential recruits.

There would have been an interesting project on how capoeira and Savate can or cannot be taught online, but one that held no attraction for me.

Ethnography online has flourished for over twenty years. Boellstorff's (2008) use of an avatar to do participant observation in the early years of Second Life was an excellent pioneering project. Hine (2021) is an overview of the achievements and potential of online ethnography. The pandemic certainly provided an opportunity to do online ethnography of capoeira and Savate teaching, and I could have done that. However, I decided that I did not want to begin such an investigation myself. I had more data than I would ever 'use' and am extremely technology adverse. I hope others would do such studies. So, I left my friends, teachers, students and their children as abruptly as any of the contributors in the literature. None of the strategies to maximise the insights and data-collection strategies advocated in methods books (Delamont and Atkinson 2021: ch. 9) or the sparse autobiographical literature were used.

I hate having no field site to go to. The reporter's notebooks I use in the field (all 220 of them) are in a colleague's office in the university. Number 221 is in the fieldwork bag, and perhaps the most pathetically unused ones are labelled up to 241 and sit awaiting use on a bookshelf. I use three sets of notebooks: reporter's notebooks in the field, A4 books to write up the material and A5 books to keep an out-of-the-field diary. There is a stock of the A4 books (six) and the A5 books (three), also labelled and ready to use. There were several events scheduled for the summer of 2020. It would have been M. Poncianinho's twentieth anniversary festival in London with about thirty teachers from all over the world, many of them old friends of mine whom I was looking forward to seeing, and its flyer sits forlornly in the out-of-the-field diary.

My written-up fieldnotes (Book 115) stop dead on 5 March 2020, with the now poignant entry 'I'm glad I went'. I did not go to a class in Cardiff on 14 March 2020 – which, it transpired, was the last to take place before the community hall closed – because, as I wrote in my out-of-the-field diary (Vol. 15), 'I felt very tired from the picket line': we were on strike that week. That was the end of 'live' classes in Cardiff for sixteen months. The 'out of the field' diary continues to the present day with rather sparse entries, mostly about the progress, or lack of it, of publications. My fieldwork bag is still packed to go under the stairs, up-to-the-attic ready: the only thing removed is the water bottle. Otherwise, the pens, notebooks, plasters and small towel are where I left them in March 2020 and the drawer of capoeira T-shirts that I wear to the classes and festivals sits unopened.

Coffey (1999: 106) points out that the advice offered by Lofland and Lofland (1995: 63) implies that 'ending fieldwork is a necessity to be skilfully managed'. Yet for Coffey it is better seen as 'an experience to be lived through' and 'closing a chapter of one's life'. I did not want to close that chapter and had no desire to live through the leaving experience.

Rob's story

As with Sara, I confess to being rather bad at leaving the field. In the decade since my PhD, I have 'only' completed one ethnographic project: a participatory field study of outreach work with the street homeless. My current project, another fully participatory study of the work of a mountain rescue team, very nearly ended up being the focus of this tale when there was a serious fire at the team's base just eleven months on from my joining in 2017. I will never forget receiving the SARCALL[1] text message that read:

> CBMRT FT Base is on fire, 501 and 503 on scene with fire service. Base will be non-Ops for foreseeable and any calls will be handed to other teams 501 21:28hrs 25-11-17.

Through a truly amazing amount of hard work on the part of team members and the generosity of the public, fellow mountain rescue teams, charities and other organisations, the team stayed operational. I am very pleased to say that at the time of writing, almost four years to the month, the team has moved back into the completely restored and refurbished base. I often use this event as a cautionary tale to students of the need to get as much data gathered when you can. You never know when your project might go up in smoke. My fieldwork with the mountain rescue team, happily, continues. Regarding our above comment about intellectual comfort zones, I continue to learn about both the skill of mountain rescue work and the organisation of the team's activities as both member *and* researcher. Indeed, I am currently training to be a Party Leader, which will yield greater insights into the coordination of the work of call-outs (my main research interest).

The project I have left, through choice, was that study of outreach work with the street homeless (see Smith 2011; Hall and Smith 2015; Smith and Hall 2018). It was my first research post-PhD, and the formal fieldwork was scheduled to last for eighteen months. The project, initially, was concerned with 'urban patrols' broadly conceived, but it was with the outreach workers of Cardiff's Housing and Neighbourhood Renewal (HANR) team (acting as a volunteer on the various charitable outreach projects they support) that I spent most time with and, subsequently, became most attached to. The field engagement and writing ended up lasting another six years and I became one the longest-serving volunteers on the Breakfast Run project. Those seven years or so were, of course, experienced in different rhythms and intensities; however, for most of the weeks in that time frame I would spend at least one, and often more, of my mornings on the Breakfast Run.

This involved arriving early at a temporary accommodation hostel in the city centre to help the hostel volunteers prepare and pack up the HANR team's van – or often one of our cars – with various breakfast items, spare clothes if available and an often random assortment of donated items, before heading out to meet up with and search for the rough-sleeping homeless; looking at the city centre through 'outreach eyes' (Hall and Smith 2017).

Reading Sara's account above, I recognise that in being taught ethnography by her and Paul Atkinson, I also received the notion that an ethnographer should always be doing fieldwork. That sensibility does not sit so well with funding schedules, submission deadlines and Gant charts, but seems the right attitude for people who set out to be students of society. More specifically, a principal methodological reason for my *not* leaving the field was the sense in which I wanted to feel I was writing about something 'live' and not as a historical document coloured by time and memory and shifting interpretations of what was going on during those mornings spent on city-centre streets. One can continually discover aspects of ethnographic naivety when what you think you might commit to paper is confronted by what is being done in practice. Whatever it was that I was thinking about and writing at the time, I could, casually, drop it into conversation on one of the early mornings and receive confirmation, further examples in the form of past stories or, indeed, a rejection of the theory in progress. Still, I did become aware of a certain routineness to the fieldwork and was aware, perhaps, of moments far ahead of the end of the seven years in which I was seeing things *as* an outreach worker. I vividly recall one morning where I volunteered a complaint about a client describing their situation – something I had not done previously – and the outreach worker nodded and said 'yeah, exactly'. Still, fieldwork is often, somehow, not only about the recording and gathering of data but about how the experiences of fieldwork permeate your everyday life and, as Sara notes in a different way, your sense of self. I gained a lot more from those mornings than fieldnotes and reflection. At last, however, it really was time to go. Not due to anything as grand as my having reached 'saturation' point but because I wanted to begin another project and join the mountain rescue team.

In December 2016 I told the charity, and the outreach team, of my intentions to stop volunteering on the project at Christmas. It was done with no little regret, although also with a little relief at no longer having to set the alarm clock quite so early. I had developed lasting relationships with the HANR team – particularly with Dennis, Charlie, Sue (who left before I did) and the legendary Jeff, who sadly passed away in 2018. In January I received a call from the team saying that the Breakfast Run was short-staffed the next day, and could I help out? When you become a member of a setting in the full sense, there are sometimes good practical reasons why leaving can be hard. I said, yes, of course, and arrived at the hostel like the many times before. On the round that morning we stopped at a back-street church in the city centre. There, with a couple of other regulars, was Dave. Dave was a guy that I'd known for a long time: a quiet bloke who'd always kept himself to himself – *never* let on where he was sleeping – always relatively well turned out, clean shaven, never any trouble. He had been picked by the team as someone who would benefit from moving into accommodation rather than a hostel and was fast-tracked into a flat. The team had made the right call, and Dave was doing well, had even got himself a job interview. The problems started when Dave was late for the interview, through no fault of his own; 'Given the wrong time, wasn't I?' He called and rushed into town, but was late. And that was it. Missing the interview went against him and things piled up from there, to the extent that he'd lost his flat. Dave had been 'fucked over' by the system. Hence his reappearance at the Breakfast Run.

I'd always got on with Dave and he squinted and smiled in between drags on his rollie when he saw me getting out of the van. 'Haven't seen you about in a while, Rob.' 'Yeah mate, I've been busy with uni and that. You know, work stuff. I've had to stop coming out.' 'Right,' said Dave, with a lingering look. 'I knew you'd give up on us in the end.'

Dave's assessment provided a coda to my fieldwork that will probably always stay with me. The pleasure of being recognised and missed as an ethnographer on my return was strong. We often imagine our participants will miss us even if that is not the case. That was swiftly followed by the recognition that I had, indeed, been just a visitor who, much like everyone else who had shown a friendly face to Dave at some point, would eventually let him down. That is the emotional response, at least. Emotional entanglements that are considered real are real in their consequences; and some entanglements are very real indeed (see Chapter 5).

Analytically, though, and perhaps less self-indulgently on my part, that interaction is itself characteristic of outreach encounters more generally. A vulnerable and distrusting population who, nonetheless, might be seen to be out to get what they can from encounters with people who have something to offer (Rowe 1999). These kinds of personal connections have, for all parties, a Janus-faced character. They are transactions. The rough sleepers after a breakfast, a ciggie, a bump up the queue, the whereabouts of that guy that owes them money. The outreach workers are also doing their *job*: after information – where *is* Chris sleeping at the moment, he's got to get to that appointment at the clinic or he'll lose his spot on the Shoreline programme – after results of some sort, and under pressure from management to justify their slow, patient, street work. And let us be clear, the ethnographer's presence in such settings is transactional too – there to join in, become 'one of the team', but also to gain something more 'through one sneaky means or another' (Goffman 1989): the right to *be there*, relatively unnoticed, making observations, producing data, to get that project done. As I have suggested elsewhere, such entanglements are the very stuff of the politics of care and kindness (Hall and Smith 2015), and leaving those entanglements can be hard for all parties. Exits are, of course, hardest of all for those who really could do with leaving those streets behind.

Beyond the emotional ties, and genuinely missing those mornings in Cardiff city centre, I often reflect on the extent to which I ever really 'left' that field. What would 'leaving' actually mean? In Chapter 9, Andrew P. Carlin notes a similar revealing of an ever-present field after his education in street order from a *Big Issue* seller. After spending so many hours driving and walking, slowly, around Cardiff, on the look-out for rough sleepers and the traces of their presence and movement, I have never looked at city space in the same way (Hall and Smith 2017). When I walk through a city centre, I routinely do so with 'outreach eyes', noting flattened cardboard behind phone kiosks, sleeping bags stashed in bushes, the small orange caps from needles, balled tin foil and myriad other traces of a field I have never really left.

The view from the exit(s)

The remainder of the book is organised in to four intersecting and sometimes overlapping parts that deal with distinct aspects of leaving the field that we have touched upon in our introduction. We do not introduce the chapters in detail, here. The authors do an admirable job of introducing their own exits. And as we have taken ample room here to do the work of reviewing the extant exit literature, they are free to get straight into their tales of leaving the field. We see this very much as a book that one might 'dip into'. Each story can be read in its own terms, although we would like to think that a cover-to-cover reading would be a rewarding experience. Still, for the sake of orienting the reader to what is in store, we outline the parts and the chapters that they gather.

Part I: Entanglements and im/perfect exits

Part I considers ethnographic exits in relation to the entanglements with and within the field. The five chapters, in different ways, take forward some of the issues that Rob's story raised, and consider the relations and relationships that colour the process and experience of exiting. As already noted, much has been written of field relations in the midst of fieldwork, but what of those relations when the research comes to end, whether as planned or abruptly in 'good' and 'bad' exits. These chapters, like the others in the book, display an openness and honesty in considering the ways in which various exits were done. We learn that very rarely is an exit 'done tidy' for all parties. Indeed, we begin the collection with a chapter on making the best of imperfect exits.

In Chapter 1, Allan and Cole reflect on two separate instances of leaving the field – from an elite single-sex girls' school and a comprehensive secondary school, both in the UK – in which various conditions provided challenges for them to meet the ethical imperative of ending projects 'well'. Through the discussion of those challenges, they demonstrate how the sometimes 'messy' business of exiting sheds light on the inadequacy of prescription of what a 'good' exit might mean. Returning to the Michailova et al. (2014) paper discussed above, Allan and Cole, in considering how they navigated their imperfect exit, propose the metaphor of 'labyrinth' in the place of 'black hole'. As the authors discuss, the metaphor of the labyrinth recognises the myriad challenges of exiting while also acknowledging that there *is* the possibility of an exit. The trick is attending to the multiple pathways encountered on the journey.

In Chapter 2, Sally Campbell Galman presents a graphic Shakespearian meditation (literally, as you will see) on 'bad exits'. She reflects on 'leaving' as a social and cultural phenomenon more generally, and critiques the 'leaving the party' model – which suggests that there are 'correct' and 'incorrect' ways to leave the field, and thus 'good' and 'bad' exits – and the 'break up' model that implies exiting must be painful. Drawing from her 'insider' ethnography of the experiences of trans and gender-diverse children in schools and communities, Galman reflects on the difficulties that coloured her own messy 'bad' exit in which multiple categories and relations must be accounted for. Galman concludes that, despite their awkwardness, 'bad exits' that honour complexity are ultimately humanising, and might occasion a revisit to the way we have understood leaving.

In Chapter 3, Alex McInch and Harry Bowles reflect on their different experiences of leaving different field sites – a working-class secondary school and a cricket academy – and the ways in which plans made and not made can never provide for the unpredictability of how ethnographic exits will play out. The chapter starts with McInch's story. Despite his planning for the continuation of relationships and, indeed, hoping to follow the journeys of students – and 'Wesley' in particular – the exit, for McInch, was unexpectedly abrupt and final. The project just ended. Still, McInch reflects on how this experience provides insights into those of teachers and their dealings with the cycle of compulsory education. The chapter then continues with Bowles's account of his exit from the field which foreshadows some of the issues described in the following section. In a direct contrast with McInch's cutting of ties, Bowles reflects on friendships in the field and, through his relationship with the coach, some of the obligations and commitments that can find the researcher bound to the field in unexpected ways. Together the authors make a case for the recognition of exiting as a relational process, coloured by care and compassion, that resists reduction and prescription.

Continuing the theme of entanglements and exits where best-laid research plans run up against the realities of (academic) life, in Chapter 4, Gareth Thomas describes his exit from fieldwork in a post-industrial town, prompted by the end of a fixed-term research contract. Thomas considers how this somewhat arbitrary exit was affectively experienced as a rupture of the emotional attachment to his informants. Significantly, he does so in relation to what he felt was a betrayal of those informants in terms of the responsibilities of the researcher. In foreshadowing some of the themes covered across the book, Thomas considers the rupturing of these field relations right through to the politics of presentation and the continued responsibility we have in writing about those whose lives we have, temporarily, joined.

Emotional ties and affective exits are further and powerfully discussed in Chapter 5 by Erin O'Connor, who provides an intimate and eloquent discussion of her own romantic attachment during her ethnography of an apprenticeship in glass blowing. O'Connor writes openly about her experience of 'love in the field' – culminating in becoming engaged to be married to Sarkis, a 'hot shop hero' – not so as to reflect on entanglements in the sterile austere manner often found in the recommendations of textbooks and ethics committees but, rather, to reflect on the intersections of the body, of gender and of the b(l)ooming of love in fieldwork. O'Connor encourages us to think about 'the field' as something more than a bounded site, as something that is emergent and revealed, even in the act of exiting.

Part II: Troubling the field

Part II, containing five chapters, focuses specifically on how exits of different kinds trouble what we talk about when we talk about leaving the field. These chapters draw from a wide range of ethnographic approaches – from anthropology to ethnomethodology – but the commonality is that the authors' experiences of exiting have led them to question, in related yet distinct ways, what 'the field' actually is, how the 'field' might be far more unstable than is often recognised and, given some of what has been already said about entanglements, how and whether one can actually be said to 'leave'.

Part II begins with Chapter 6 in which Jessica Nina Lester and Allison Anders reflect on the ending of a four-year community-based ethnography. Considering the aspects of the field that were easy to leave behind and those that lingered after the completion of the project, the authors point to how memory, a key aspect of the ethnographic endeavour, and continued communications with informants mean that 'the field' is not so readily left behind. Adopting a post-critical orientation to ethnography, the authors consider the partial and positioned character of power and the contrasting trajectories of the researchers and Burundian refugees with whom they are still in contact today. The authors point to a key aspect that many ethnographers will recognise – the ways in which memory is a form of commitment to the field and a site in which the meaningful connections with those whose lives we join live on.

This theme is taken up in Chapter 7 by Dawn Mannay, who considers similar extensions of the field which complicate the notions of exiting, which, in turn, provokes a reflection on the boundedness of 'the field' as a concept. Mannay considers how the researcher's status within a community of practice – and their movements from periphery to centre – can produce a sense of permanency within an expanded field of relations. She draws on notions of *déjà vu* ('already seen') and *jamais vu* (never seen) to consider how experiences and phenomena are recurrent not only in fieldwork but historically too, as well as how the supressed aspects of self and status can jar the fieldwork present. Mannay considers the non-linear, 'messy' and recursive stages of ethnographic research and, like the previous chapter, how community and advocacy research can mean an enduring tie to a shifting and expanding field.

In Chapter 8, Janean Robinson, Barry Down and John Smyth also consider the significance of echoes of informants' voices after fieldwork had formally concluded. They propose that the key component of critical ethnography is to adopt an attitude that ever really leaving the field is an impossibility, and point to how maintaining and respecting such contacts can be a subversive and radical act in the context of neoliberal society. The authors recount their tracing of the fortunes of two students after they sent a text message to participants at the formal conclusion of the research project. Tracing the experiences of the students is a moral and political commitment that, the authors argue, is central to raising consciousness, encouraging activism and providing for spaces that allow informants to 'speak back' to dominant discourses through the recirculation of their experiences.

Chapter 9, by Andrew P. Carlin, considers an expanded field that one cannot leave from an alternative, phenomenological perspective. Carlin recounts his accessing a particular phenomenal field in and through an encounter with 'Tommy' the *Big Issue* seller (a magazine sold by homeless and vulnerably housed individuals on the streets of UK cities). Taking the pavement as a research site, Carlin draws on ethnomethodological studies of public space to consider how the phenomenal field of visible poverty is accomplished in practice and, once seen properly, constitutes a recurrent field without exit. Carlin also presents a critical engagement with previous discussions of entering and leaving 'the field' as presented in urban ethnography, before going on to discuss his practical tuition in 'pavement culture' from Tommy. Carlin describes the category work achieved by Tommy in maximising his success in selling magazines and, also, managing his own presence within the pavement order. With pavement culture viewed in this way, which recognises how begging practices are adapted to, and exploit, pavement culture, the field can never be left behind.

This part, on troubling the notion through an attention to possible and impossible exits, closes with Chapter 10 and George Jennings' reflections on the difficulties of 'leaving' when the research is participatory and concerned with an embodied practice. In describing the 'marriage to skilful practice' encompassed in his learning of the martial art of Wing Chun Kung Fu, Jennings considers both the work of learning a deeply embodied skill and how the field thus becomes written into bones and muscle. Reflecting on various fieldwork journeys and, importantly, the various leavings – from family, from informants, from friends – that a fieldwork career entails, the chapter describes a process of becoming and how, thereafter, the field – considered in relation to the concept of embodied habitus – thus travels *with* the researcher.

Part III: Intermissions and returns

As readers will well know, what is often referred to as 'the fieldwork period' is never a linear and uniform passing of time. All fieldwork is characterised by different rhythms, patterns and intensities. Even the most immersive projects are punctuated by interruptions, withdrawals and returns. In Part III, four chapters each consider a distinct aspect of this patterned character of engagements with the field, while all focus upon drawing out methodological insights into temporary exits and returns, and all those experiences in the field that are somewhere in between.

Starting with Chapter 11, Alice Corble draws on her fieldwork in a public library affected by the austerity policies and cuts to public spending in the UK. In her chapter, Corble discusses recursive returns to the field as bound up with her own shifting understandings of self and identity. She reflects on the entanglements which made it difficult for her to leave the field and how the 'messy' business of tracking back and forth can provide insights into 'the field' as well as ethnographic and activist practice. Corble concludes with three lessons that are very much aligned with our notion of the 'active exit' and researcher discussed earlier, and asks readers to attend, analytically, to 'fuzziness', to embrace contradictions and to consider the implications of praxis-oriented research for the understanding of power relations.

In Chapter 12, Joe Williams draws on his experience of a fieldwork intermission. Describing something of his ethnography of outreach workers in New York City, Williams reflects on his dissatisfaction arising from presenting his work at a conference on his return home. Drawing on an ethnomethodological sensibility, he considers how the category 'homeless' is constructed in particular ways within social science discourses. Although, of course, the term is contested, Williams shows how his temporary withdrawal from the field, and the confrontation with academic renderings of the category, acted as an impulse, upon his return to the streets of Manhattan, to pay attention to the ways in which the category is accomplished within, and meaningful for, the work of outreach.

Chapter 13, by Andrew Clark and Sarah Campbell, reflects on research with people living with dementia. The project was itself concerned with place and the experience thereof, understood as a relation of the geographic, the temporal and the social. In exploring 'the field' in this way, through the experience of those living with dementia, their project not only interrogates the constitution of 'the field' conceptually but also challenges received notions of entering and leaving by switching the perspective from which many accounts are constructed. Rather than centring the troubles of the research and researchers, the authors consider the physical and cognitive difficulties faced by the participants themselves in accessing 'the field' constructed by the research. The authors draw on the experience of this research – for all parties – to recast 'the field' as a cognitive as well as physical location for which 'leaving' can take on multiple dimensions.

David Calvey, in Chapter 14, reflects on his 'constant apprenticeship' in martial arts and, specifically, Jeet Kune Do. In a related yet distinct manner to Chapter 10, Calvey also questions the nature of 'the field' as something visited, accessed and left in terms of geographic setting. Where Jennings in Chapter 10 describes the difficulties in moving between different forms of martial art, in Chapter 14 Calvey's tale presents a very different sense in which fields can be hard to leave, particularly when they intersect. In this case, Calvey reflects on intermissions in fieldwork, and the presence of one field within another in the overlapping fields of martial arts and the world of bouncing. Calvey considers the work of managing the ways that martial arts travelled with him into his covert research. Chapter 14, then, considers the dynamics of these intersecting fields, one that he has left and one that he will not leave, and the complexity of the entanglements, revisits and regrets that colour what he sees as an unavoidable part of the craft of ethnography; an aspect not to be reduced or sanitised in accounts of 'clean' exits.

Part IV: Returns, responsibilities and representations after 'leaving'

The fourth and final part contains three chapters that focus upon a theme hinted at across the collection – the post-fieldwork 'consequences' of research in terms of the writing and publication of observations about those we have spent time with 'in the field' (see Brettell 1996). The chapters here all reflect upon the tensions that exist between the role of participant (to a greater or lesser extent) observation and that of the analyst and author.

Chapter 15, by Daniel Burrows, offers a cautionary tale in relation to the practice of 'respondent validation'. On returning to the field, Burrows reflects upon some of the tensions inherent in returning as an author, as opposed to fieldworker, and how – in a similar manner to the observations in other chapters – this raised issues relating to his sense of self. Drawing on notions of the situated self, Burrows describes some of the discomfort in returning to the hospital social work practitioners he had studied. Shifting from his claims to be 'one of them' to making critical authorial claims *about them* and their practice occasioned, for Burrows, not only disquiet but a deep questioning of the ethicality of his own practice. Burrows' reflection raises significant questions relating to the ownership of ethnographic representations and, indeed, tensions between the 'field self' and 'author self'.

In Chapter 16 Daniel Newman continues this reflection upon the tensions many researchers will have faced in producing a necessarily critical analysis of a community they have entered and embraced. Returning to the themes of entanglement considered in Part I, Newman describes the challenge of balancing ethnographic honesty with the pressures of friendly interpersonal relations with informants (who have, of course, stopped being 'just informants' by that point). He discusses friendship as a methodology which, itself, raises questions of situated ethics when, for example, an informant indicates that they are talking 'off the record'. Beyond the frustrations of not being able to report the 'juicy' data, Newman reflects upon how a key aspect of the research was itself about the tension between what the lawyers he studied said they did, and what he observed them doing in practice. In reflecting on leaving the field, and the impact of writing about the lawyers in a critical light, Newman talks about the need to manage a lasting negative potential of his book, but also reflects upon the need for novice researchers to be aware of, and prepare for, the eventuality of writing critical things about the people they have grown close to.

Chapter 17, the final chapter of the collection, returns to the theme of lingering regret (that has been peppered across the collection, in different ways) in Ashley Rogers' self-reflexive analysis of leaving her Bolivian field site and returning 'home'. Rogers openly describes some of the anxiety related to the responsibility of writing up the fieldwork; this is a familiar aspect for many, perhaps, yet Rogers makes a compelling case for the additional challenges posed by conducting (doctoral) fieldwork abroad. She offers insights into how to deal with that responsibility – and do the 'write thing' – through the use of a reflexive diary, drawing and the importance of reading the self-reflexive texts of other academics and ethnographers. Indeed, in connecting with previous chapters (particularly Chapters 6, 7 and 8), Rogers makes a case for this kind of reflection serving as a means of critical engagement with and resistance to the confines of current research regimes.

Coda

We very much hope that the evocative tales of leaving the field gathered in this collection provoke a greater attention to the politics, difficulties and professional myths of ethnographic exiting. The reader expecting to come away from the book with a list of top ten tips for leaving the field 'tidy'[2] may well be disappointed. We do, however, trust that the same reader will gain a good deal from the candour with which the chapters gathered here describe and discuss 'good' and 'bad' exits, the priority and (im)possibility of planning for exits, as well as an appreciation of the various responses to our leaving – disappointment, betrayal, acceptance, indifference – of those whom we have spent time with. The insights generated should not, as we have stressed earlier, be reduced to being only 'tales' about leaving, returning and representing the worlds we study but, rather, should be recognised as raising key questions about what it is to do ethnography as a lived, social, embodied, messy, contingent practice. In this sense, exits are just as significant as the more often reported moments of 'getting in', of accessing the field and being accepted in one way or another by 'the locals'. One might even argue that, given that the exit (in whatever form) from the field marks the culmination of the fieldwork project, it is exactly at this point where field relations, research ethics, rights, responsibilities, obligations and commitments, triumphs and regrets are most visible and therefore most amenable to analysis. *That* is why we make the argument here for the exit – good, bad, planned or forced – to be experienced *actively*. Indeed, as the chapters highlight, thinking about exits in this active way might well provide a space for a broader critical reflection and engagement with our responsibilities and obligations, our priorities and pressures, and just how we go about, conceive of and write about our ethnographic research. In this sense, attending to leaving the field analytically might find us not straightforwardly turning our back on the field as some pristine object but, rather, coming back to where we started off and questioning the nature of 'the field' as it is constructed and reconstructed in methodological and theoretical writings. There is a still a good deal of work to be done in shifting the legacy of the colonial gaze and privilege of different forms in ethnographic work. Notions of insider/outsider, of researcher and 'native', of research 'subjects', of centre and margin and of 'the field' as somewhere one takes oneself to and then leaves to 'return' to 'normal life' (the normal life of an academic, that is), require interrogation and unsettling. We think that attending to the moments of leaving the field, and the reverberations and hauntings and repercussions of those moves, is one way forward. We hope that this book contributes to such a project.

Acknowledgements

Mrs R.B. Jones did some of the administration, and word-processed the bulk of this introduction, for which we are grateful. We are very grateful to the scholars who refereed the chapters for us: P. Atkinson, K. Chen, B. Fincham, M. Hammersley, L. Murray, J. Newman, J. Rainford, T. Rapley, L. Russell, D. Spencer, M. Travers, G. Walford and M.R.M. Ward. We had offers of help from several other colleagues whose expertise we drew on, especially Amanda Coffey, Nicola de Martini Ugolotti, Tia De Nora, Dariusz Dzienwanski, Teresa Gowan, Craig Owen and Darin Weinberg.

We both benefit from our membership of the Cardiff Ethnography Group, a source of intellectual stimulation, useful criticism and sometimes cake. Founded in 1974, it remains a flexible and robust scholarly community.

Notes

1 SARCALL is the multi-agency Search and Rescue (SAR) Incident Management Platform used by UK Mountain Rescue, Lowland Search, Cave Rescue Teams (https://sarcall.com).
2 A South Walian expression we are fond of. 'Doing it tidy' is something of an organising principle for work that can be described under the (slightly tongue in cheek) header of 'the Cardiff School of Ethnography' (see Smith 2017).

References

Altheide, D. (1980). 'Leaving the newsroom', in W.B. Shaffir, R.A. Stebbins and A. Turowetz (eds) *Fieldwork Experience*. pp. 301–310. New York: St Martin's Press.
Atkinson, P.A. (1996). *Sociological Readings and Rereadings*. Aldershot: Ashgate.
Atkinson, P.A. (2017). *Thinking Ethnographically*. London: SAGE.

Atkinson, P.A., Coffey, A. and Delamont, S. (2003). *Key Themes in Qualitative Research*. Walnut Creek, CA: AltaMira Press.
Barley, N. (1983). *The Innocent Anthropologist*. Harmondsworth, Middlesex: Penguin.
Belousov, K., Horlick-Jones, T., Bloor, M., Gilinskiy, Y., Golbert, V., Kostikovsky, Y. Levi, M. and Pentsov, D. (2007). 'Any port in a storm'. *Qualitative Research*. 7(1): 155–175.
Blackwood, E. (1995). 'Falling in love with an-Other lesbian: Reflections on identity in fieldwork', in D. Kulick and M. Willson (eds) *Taboo: Sex, Identity and Erotic Subjectivity in Fieldwork*. pp. 51–75. London: Routledge.
Boellstorff, T. (2008). *Coming of Age in Second Life*. Princeton, NJ: Princeton University Press.
Bosse, J. (2015). *Becoming Beautiful*. Urbana, IL: University of Illinois Press.
Brettell, C. (ed.) (1996). *When They Read What We Write*. Westport, CT: Bergin and Garvey.
Burawoy, M. (2003). 'Revisits: An outline of a theory of reflexive ethnography'. *American Sociological Review*, 68(5): 645–679.
Campbell Galman, S.C. (2019). 'Flat caps and dengue fever', in R.J. Smith and S. Delamont (eds) *The Lost Ethnographies*. pp. 95–108. Bingley: Emerald.
Cannon, S. (1992). 'Reflections on fieldwork in stressful situations', in R.G. Burgess (ed.) *Learning about Fieldwork (Studies in Qualitative Methodology, Volume 3)*. pp. 147–182. Greenwich, CT: JAI Press.
Coffey, A. (1999). *The Ethnographic Self*. London: SAGE.
Coffey, A. (2018). *Doing Ethnography*. London: SAGE.
Davis, K. (2015). *Dancing Tango*. New York: New York University Press.
Delamont, S. (2012). 'Milkshakes and convertibles', in N. Denzin (ed.) *Studies in Symbolic Interactionism* (Vol. 39). Bingley: Emerald.
Delamont, S. (2016a). *Fieldwork in Educational Settings*. London: Routledge.
Delamont, S. (2016b). 'Time to kill the witch?' in M. Ward (ed.) *Gender Identity and Research Relationships*. pp. 3–20. Bingley: Emerald.
Delamont, S. and Atkinson, P.A. (2021). *Ethnographic Engagements: Encounters with the Familiar and the Strange*. London: Routledge.
Delamont, S., Stephens, N. and Campos, C.R. (2017). *Embodying Brazil: An Ethnography of Diasporic Capoeira*. New York: Routledge.
Ellis, C. (1986). *Fisher Folk*. Lexington, KY. University Press of Kentucky.
Fielding, N. (2004). 'Getting the most from archived qualitative data: Epistemological, practical and professional obstacles'. *International Journal of Social Research Methodology*. 7(1): 97–104.
Fine, G.A. (1983). *Shared Fantasy*. Chicago, IL: University of Chicago Press.
Fine, G.A. (1985). 'Occupational aesthetics'. *Urban Life*. 14(1): 3–32.
Fine, G.A. (1987). *With the Boys*. Chicago, IL: University of Chicago Press.
Fine, G.A. (1996). *Kitchens*. Berkeley, CA. University of California Press.
Fine, G.A. (1998). *Morel Tales*. Cambridge, MA: Harvard University Press.
Fine, G.A. (2013). 'Sticky cultures'. *Cultural Sociology*. 2(4): 395–414.
Fine, G.A. (2018). *Talking Art*. Chicago, IL: University of Chicago Press.

Gallmeier, C.P. (1991). 'Leaving, revisiting and staying in touch', in W.B. Shaffir and R.A. Stebbins (eds) *Experiencing Fieldwork*. pp. 224–231. Newbury Park, CA: SAGE.

Glaser, B.G. and Strauss, A.L. (1967). *The Discovery of Grounded Theory*. Chicago, IL: Aldine.

Goffman, A. (2014). *On the Run*. Chicago, IL: University of Chicago Press.

Goffman, E. (1989). 'On fieldwork'. *Journal of Contemporary Ethnography*. 18(2): 123–132.

Hall, T. and Smith, R.J. (2015). 'Care and repair and the politics of urban kindness'. *Sociology*. 49(1): 3–18.

Hall, T. and Smith, R.J. (2017). 'Seeing the need: Urban outreach as sensory walking', in C. Bates and A. Rhys-Taylor (eds) *Walking through Social Research*. pp. 39–53. Abingdon: Routledge.

Hammersley, M. and Atkinson, P. (2019). *Ethnography: Principles in Practice*. (4th edn). Abingdon: Routledge.

Hine, C. (2021). 'Ethnographies in online environments', in P. Atkinson, A. Cernat, S. Delamont, J. Sakshaug and R. Williams (eds) *Sage Research Methods Foundations* (Vol. 3). pp. 1491–1500. London: SAGE.

Hobbs, D. (1988). *Doing the Business*. London: Open University Press.

Iversen R. (2009). '"Getting out" in ethnography'. *Qualitative Social Work*. 8(1): 1–96.

Iversen, R. (2019). 'Leaving the field', in P. Atkinson, A. Cernat, S. Delamont, J. Sakshaug and R. Williams (eds) *Sage Research Methods Foundations* (Vol. 6). pp. 2749–2755. London: SAGE.

Kaplan, I. (1991). 'Gone fishing, back later', in W.B. Shaffir and R.A. Stebbins (eds) *Experiencing Fieldwork*. pp. 232–237. Newbury Park, CA: SAGE.

Keiser, L. (1969). *The Vice Lords*. New York: Holt, Rinehart and Winston.

Letkemann, P. (1980). 'Crime as work: Leaving the field', in W.B. Shaffir, R.A. Stebbins, A. Turowetz (eds) *Fieldwork Experience*. pp. 292–301. New York: St Martin's Press.

Lofland, J. and Lofland, L.H. (1995). *Analysing Social Settings: A Guide to Qualitative Observation and Analysis*. Belmont, CA: Wadsworth.

Maines, D., Shaffir, W.B., Haas, R.A. and Turowetz, A. (1980). 'Leaving the field in ethnography', in W.B. Shaffir, R.A. Stebbins and A. Turowetz (eds) *Fieldwork Experience*. pp. 261–80. New York: St Martin's Press.

Michailova, S., Piekkari, R., Plakoyiannai, E., Ritvala, T., Mihailova, I. and Salmi, A. (2014). 'Breaking the silence about exiting fieldwork: A relational approach and its implications for theorizing'. *Academy of Management Review*. 39(2): 138–151.

Oakley, A. (2016). 'Interviewing women again: Power, time and the gift'. *Sociology*, 50(1): 195–213.

Roadburg, A. (1980). 'Breaking relationships with research subjects: Some problems and suggestions', in W.B. Shaffir, R.A. Stebbins, and A. Turowetz (eds), *Fieldwork Experience: Qualitative Approaches to Social Research*. pp. 281–291. New York, NY: St. Martin's Press.

Rowe, M. (1999). *Crossing the Border: Encounters between Homeless People and Outreach Workers*. California: University of California Press.

Scott, S. (2019). 'Researching underwater', in R.J. Smith and S. Delamont (eds) *The Lost Ethnographies*. pp. 79–84. Bingley: Emerald.
Shaffir, W.B. and Stebbins, R.A. (eds) (1991) *Experiencing Fieldwork*. Newbury Park, CA: SAGE.
Shaffir, W.B., Stebbins, R.A. and Turowetz, A. (eds) (1980). *Fieldwork Experience*. New York: St Martin's Press.
Sinha, S. and Back, L. (2014). 'Making methods sociable: Dialogue, ethics and authorship in qualitative research'. *Qualitative Research*. 14(4): 473–487.
Smith, R.J. (2011). 'Goffman's interaction order at the margins: Stigma, role, and normalization in the outreach encounter'. *Symbolic Interaction*. 34(3): 357–376.
Smith, R.J. (2017). 'Doing it tidy: The open exploratory spirit and methodological engagement in recent Cardiff ethnographies', in S. Delamont, P. Atkinson, A. Coffey, A. and R.J. Smith (2019). *Lo spirito esplorativo libero La scuola di Etnografia di Cardiff 1974–2017*, Traduzione di Giuseppina Cersosimo, Chiuso in stampa nel mese di marzo 2019. pp. 68–81. Reggio Calabria: Presso Creative 3.0.
Smith, R.J. (2022). 'Fieldwork, participation, and unique-adequacy-in-action'. *Qualitative Research*. https://doi.org/10.1177/14687941221132955.
Smith, R.J. and Delamont, S. (eds) (2019) *The Lost Ethnographies*. Bingley: Emerald.
Smith, R.J. and Hall, T. (2018). 'Everyday territories: Homelessness, outreach work and city space'. *The British Journal of Sociology*. 69(2): 372–390.
Smith, R.J., Ablitt, J., Dahl, P., Hoyland, S., Jimenez, P., John, Z., Long, F., Sheehan, L. and Williams, J. (2020). 'Ethnography and the new normal. Observational studies, analytic orientations, and practical ethics'. *Etnografia e ricerca qualitative*. 13(2), 195–205.
Snow, D.A. (1980). 'The disengagement process: A neglected problem in participant observation research'. *Qualitative Sociology*. 3: 100–122.
Southwood, J.V. and Delamont, S. (2018a). 'Tales of a tireur'. *Martial Arts Studies*. 5: 72–83.
Southwood, J.V. and Delamont, S. (2018b). 'The tireur as teacher', in C. Crosby (ed.) *Context and Contingency: Research in Sport Coaching Pedagogy*. pp. 28–33. Cambridge, UK: Cambridge Scholars Press.
Southwood, J.V. and Delamont, S. (2018c). 'A very unstatic sport: An ethnographic study of British Savate classes'. *Societies*. 8(4): 122–136.
Stebbins, R.B. (1991). 'Do we ever leave the field? In: W.B. Shaffir and R. Stebbins (eds) *Experiencing Fieldwork*. pp. 248–255. Newbury Park, CA: SAGE.
Taylor, S.J. (1991). 'Leaving the field: Research, relationships, and responsibilities', in W.B. Shaffir and R. Stebbins (eds) *Experiencing Fieldwork*. pp 238–247. Newbury Park, CA: SAGE.
Van Maanen, J. (2011). *Tales of the Field*. Second edition. Chicago, IL: University of Chicago Press.
Vannini, P. (2012). *Ferry Tales: Mobility, Place, and Time on Canada's West Coast*. New York, NY: Routledge.
Wolf, D.R. (1991). 'High risk methodology: Reflections on leaving an outlaw society', in W.B. Shaffir and R.A. Stebbins (eds) *Experiencing Fieldwork*. pp. 211–223. Newbury Park, CA: SAGE.

Part I

Entanglements and im/perfect exits

1

Finishing fieldwork in less than perfect circumstances: lessons learned in 'labyrinth' exiting

Alexandra Allan and Sarah Cole

Introduction

Few of the accounts focusing on researchers' experiences of leaving the field have specifically focused on problematic exits. Michailova et al.'s (2014) work on 'black hole exiting' is an exception to this. These authors use this phrase to refer to unusually abrupt and unanticipated endings to fieldwork. They suggest that exiting the field can happen in a variety of ways and they liken this to the experience of actors leaving the stage – sometimes these exits are executed through a simple 'exit stage left', but other times through a more melodramatic departure. The same can be true for fieldwork exits, they say, with some comprising little more than a timely, heartfelt goodbye, but others taking a more dramatic form, perhaps involving a prolonged ending in emotionally challenging circumstances. Michailova et al's (2014) work is a powerful reminder that not everything is straightforward in the process of leaving the field.

This chapter seeks to augment this argument – to continue to redress this imbalance in the literature and to widen the methodological conversations which exist around uneasy fieldwork experiences. It does so by offering two further accounts of ethnographic fieldwork which ended in less than perfect circumstances. The first tale draws from Alexandra's experiences of exiting fieldwork in an elite, single-sex girls' school in the UK; the second from Sarah's experience of exiting a comprehensive secondary school in the UK. Both tales outline the various struggles the two researchers faced when attempting to uphold the ethical imperative to 'end well' in challenging circumstances. The tales conclude with a note on the collective lessons learned from these difficult exiting experiences and a suggestion for a new metaphor – the labyrinth – as an alternative to the existing proffer of the 'black hole'.

Alexandra's tale

Alexandra's fieldwork took place in the context of an elite, single-sex girls' school in the UK. This was a selective, fee-paying institution which catered for pupils from the ages of 3 to 18 in distinct junior, senior and sixth-form departments. The focus of her research was gender and academic achievement – it sought to explore how pupils in this unique educational context positioned themselves in relation to academic achievement and how this was constituted alongside gendered subjectivity. The project took place over two years (2003–4) and comprised two phases of study. The first phase took place in the latter half of 2003, over a six-month period from February to July, with a group of twenty-five girls who were in their final year of the junior department (aged 10–11). The second phase took place with the same group of participants over a further sixth-month period between September 2003 and February 2004. At this time the girls were in their first year of the senior department (aged 11–12).

This was an ethnographic project underpinned by strong participatory principles. The girls became involved in most aspects of the research process – data generation, analysis and dissemination. Participant observation was central to the research. Alexandra observed the girls in their everyday schooling, both inside and outside of the formal classroom, and in a range of extra-curricular activities. The research also included a series of focused group interviews and a sequence of photographic research activities. The photographic activities took place in the first phase of study and were conducted through a lunchtime club. Eight participants attended the club on a weekly basis and engaged in activities which encouraged a critical exploration of photographs and led to the creation of photographic diaries. The diaries were shared in individual photographic narrative interviews and the club culminated in a one-day photographic workshop, where the girls worked alongside a professional photographer to produce a series of their own portrait prints.

Planning to leave the field

The process of leaving the field was given a good deal of consideration in Alexandra's study, largely because of the two different phases. As well as ensuring that the fieldwork came to a suitable conclusion, she was aware that the gap between the two phases required careful consideration. As Lofland and Lofland (1995) contend, researchers returning to the field tend to be more keenly aware of exit strategies because it remains a 'live' issue. Following advice from Lofland and Lofland (1995), and Delamont (2016), Alexandra spent a good amount of time developing her exit plans with the gatekeepers at the point of negotiating her entry to the school. As the timetable for fieldwork was discussed, start points and end points were given due consideration, in order to ensure that the school was well aware of the plans ahead of time and that both the timings and the plans chimed with existing school patterns, cultures and rhythms.

The plans for leaving the field were also developed in collaboration with the participants. The girls advised on how they felt the phases of fieldwork should end; for example, suggesting that the photographic workshop should be offered in the last week of term as a celebration activity to mark the end of the first phase of the research. As the project reached its official end point, at the end of phase two, the girls decided that they wanted to work on one final project. After discussing it with their teachers, they decided to create a photographic exhibition to display some of the images from the project in the entrance to the school.

In line with earlier arrangements, it was agreed that the display would be worked on during three lunchtime sessions. During these sessions the girls looked through their images to make decisions about which to include in the display. They discussed the order, arrangement and sequencing of the images, and they created captions to accompany them on the display boards on which they were mounted. The sessions offered an interesting space in which to generate further data on the images. The girls worked together to craft understandings of which subjectivities could be represented and displayed to a wider public audience in the exhibition, and which remained 'unsayable' or 'unseeable' (Bloustien 2003). The sessions also offered space for a natural 'winding down' of the research project.

Exiting into chaos

The final session took place on the last day of Alexandra's fieldwork and it appeared to go well, with the exhibition given the final touches needed. It was towards the end of the session, as Alexandra was saying her goodbyes, that two of the girls asked if they could speak with her before she left. She stepped outside of the room, into the private space of the corridor, and the girls told her that they were concerned about some of those who had been involved in the research. They told Alexandra that one of the club members had been talking to others in the school about the research and had made hurtful comments about some of the images.

The photographs at the centre of this incident were predominantly those which had been created by participants who had wanted to represent themselves as 'girly girls' – images which had been purposefully arranged to portray a powerful and pleasurable hyper-femininity. These images often depicted the girls in short skirts, sun tops and strappy vests, deliberately laid out in front of a range of 'girly' paraphernalia (e.g. lipsticks, jewellery and handbags). It transpired that rumours had been spread throughout the school about the explicit sexual nature of these images. The girls depicted in the photographs were being bullied for being 'sluts', and for knowingly representing themselves in this way.

At this moment it struck Alexandra that the 'beautiful exit' (Michailova et al. 2014) which she had planned, and which had seemed moments away, had vanished from sight, only to be replaced by an impending sense of chaos and despair. The incident appeared to have risen out of nowhere, yet it loomed on the horizon as a huge ethical speedbump, seemingly presenting an insurmountable challenge (Weiss and Fine 2000). Alexandra's overriding concern was for the safety of the girls. The irony of this being a project which sought to explore gender diversity, and ultimately to challenge restrictive gender norms, was something not lost on her. She was troubled by the fact that the girls might have been harmed rather than helped by these explorations. Thus, this was an ending which was somewhat stricken with panic. Alexandra wondered how these issues could be addressed quickly and efficiently, particularly as they had been presented at a point when there was supposed to be no return to the field. But it was also an ending tinged with sadness, largely owing to the fact that the bonds of trust between the participants in a once collaborative project appeared to have broken down.

Fortunately, as a PhD student at the time, Alexandra was able to quickly consult her supervisors on these matters. Their advice, which she followed closely, was to respond in a way which would be in keeping with the initial research plans. This meant arranging a further meeting with the girls to discuss the matter in light of the project's original (participatory) principles and to find a way forward. It was agreed that the matter should be dealt with as other bullying matters would be, by taking it to the head of school. Those who had been vulnerable to the bullying incidents were consulted on an individual basis and asked if they felt it appropriate for their parents to be involved. The girls agreed to this, and a series of meetings were set up with parents and relevant staff members.

Alexandra attended these meetings and worked alongside those involved to discuss the matters and come up with a suitable plan to address the situation. Though the meetings were largely experienced as positive (in terms of addressing the matter quickly and in line with usual child safety practices), Alexandra left the meetings feeling despondent. In part this was because the situation still felt raw. The project was not as 'complete' as she had wished it to be and she was experiencing a tremendous amount of guilt towards her participants, feeling that previously good research relationships had been ruptured. This was particularly felt in one case where it seemed that the only logical conclusion for one family was to withdraw their daughter from the project. Alexandra's concern was also one of 'ecological heritage' (Gobo 2011) – about the future impact that these negative fieldwork experiences might have on those left in the field, or for researchers who might enter afterwards.

Alexandra's tale, then, is, one which highlights the multiple unexpected troubles that can plague a researcher at the end of their fieldwork. The happy exit which she had planned resulted in an unanticipated dramatic moment and led to considerable emotional and professional anguish for the researcher and a set of somewhat tarnished research relationships. The incident also worked to colour Alexandra's experiences for some time to come. She was uneasy about not knowing how the participants were faring in the weeks following the fieldwork, she had a dented sense of pride in her abilities as a researcher and she experienced an emerging sense of ethical paranoia about whether she had done enough to prevent the incident from happening in the first place. And all of this was compounded by the fact that she could not find accounts of exiting the field in the wider literature which were like her own, she wondered if she was the only researcher to leave the field under such a shadow.

Sarah's tale

Sarah's research sought to explore young people's views and experiences of domestic violence and abuse and to examine the role which education might play in relation to these. The initial project design included two distinct phases. The first phase was to be a pilot study with young women who had recently experienced domestic violence and abuse (DVA) in their teen relationships, while at school, and who had subsequently sought support from specialist services. Access to participants was aided by Sarah's 'insider position', having worked for DVA organisations in her previous profession. This first phase began in 2011 and involved in-depth semi-structured narrative interviews with ten young women (aged sixteen to nineteen). The project was informed by robust ethical principles drawing on a feminist ethics of care to explore and give voice to 'silenced' experiences.

The second phase of the project was deliberately designed to follow on from the first, based on the expert opinions of the first group of participants, who worked to develop and frame this subsequent phase of study. This phase was planned to take place in September 2012, as a phase of ethnographic fieldwork within two secondary school settings utilising a gendered lens through which to focus on the delivery and engagement of a healthy relationships programme aimed at preventing domestic abuse and sexual violence.

Due to the sensitive nature of the research, certain logistical barriers and the problematic confines of the robust ethical boundaries Sarah had constructed, this initial fieldwork plan became unachievable. Indeed, over time a new design was developed, based on a more opportunistic and 'magpie' approach, owing to these emerging practical and ethical considerations. At the time this was experienced by Cole with some anxiety and frustration. Her research journey had become protracted and circuitous, involving what felt like 'as many steps backward as forward' (Edmondson and McManus 2007: 1173); for example, including more than one instance of having access granted and swiftly rescinded, which, as Hey (1997) reminds us, is possible at any point in ethnographic fieldwork.

Eventually what this meant was that the first and second planned phases of study took place simultaneously and that the ethnographic fieldwork was undertaken in one UK comprehensive secondary school. It also meant that Sarah's concentration had to be doggedly fixed on the matter of gaining access. All her efforts were centred on this achievement. At the time, she was led to believe that this was something of a 'holy grail': a matter on which her burgeoning identity as a PhD researcher rested entirely. And because of this, her exit was something which was yet to be imagined.

Painful comings and goings

But Sarah's tenacity and commitment to the project did eventually pay off. An e-mail introduction from a past colleague meant that she was able to gain access to a secondary school that was noted for its 'unusual' approach, both to pastoral care and to issues relating to DVA. Sarah was excited by the possibility that this invitation would afford, especially as the shortness of time was starting to exert increasing pressure. Having established contact, Sarah was invited to attend the school the following week to start to plan this next stage of the project and the final fieldwork activity. On arrival at the school, and the 'pastoral 'hub' in which she would be based, Sarah was given a warm welcome. Following some brief introductions, she was immediately met with several questions from the animated team of six staff members present. They too began to share their experiences, including details about the contexts of children experiencing DVA in the home environment and within adolescent relationships. There was overwhelming consensus that 'domestic violence and abuse contributed to being one of the biggest factors in their work' (Pastoral hub staff discussion 2012).

Sarah spent the rest of the morning in the hub and was made to feel comfortable. The group continued to share professional experiences of teaching and the domestic violence sector and Sarah was also able to expand on her research plan. Shared values emerged in the discussion, particularly

Finishing in less than perfect circumstances 39

a keen sense of wanting to 'make a difference' in prevention work. The hub team's remit had recently extended to the facilitation of a 'healthy relationships' prevention programme. This was to replace a programme delivered by a domestic violence charity, beset by government funding cuts. Sarah was asked to share her ideas and experiences specific to the development of prevention work. A growing sense of camaraderie and mission developed from the intensity of the dialogue. The hub team's enthusiasm and backing having been garnered, the agreement for access was keenly granted by the school's head teacher later on that morning.

Thrilled at this positive news, Sarah returned to the hub to discuss a possible research timetable with the team. It was at this point that she was able to observe the hub in full swing; vibrant and lively with young people's comings and goings, including doors slamming, tears and congratulations, then the school bell, and a lull. Many of the hub team had left the main hub room to support students back to class or to a therapeutic space. Sarah then found herself alone in the hub with one of the team. This allowed for an easy alliance with the team member, who was keen to share the personal significance of their work and their desire to 'make a difference'. This intimate space also proved conducive for a personal disclosure of their own experiences of DVA as a child. With a seeming matter of urgency, the staff member shared intimate details of the devastating trauma of physical and emotional violence and the profound and lasting impact of the shame and guilt which had tainted their life. Their personal narrative was accompanied by an emotional declaration: 'I have never told anyone about this.'

Sarah was both privileged and humbled; she listened and took great care. Having worked in the field for many years, this was not the first time that she had heard such a disclosure, although she considered the impact no less poignant. She was keen to 'hold' the space, to listen and to provide a sense of security. Keenly aware of the intensity of the moment and committed to following an ethics of care, Sarah moved the conversation to provide boundaries around the emerging tale. The staff member was resolved that sharing had offered a sense of catharsis and connection, and that their experiences informed their subjectivity and their role in making a difference to children and young people with similar experiences. A knock on the door from another staff member signified an immediate end to the conversation and the need to leave the hub to go to lunch.

After lunch Sarah found herself alone in the pastoral hub with another member of staff. Once again, the conversation turned to their deep desire to make a difference in working with young people. Sarah could sense what was coming. The staff member disclosed that they were experiencing domestic violence in their current relationship, sharing that they had not disclosed this information, except to one close friend. Sarah listened intently to this heart-breaking experience until an interruption from another staff member meant that they could talk no more, apart from to make a promise to talk again when time permitted. This somewhat intense, but positive, day ended with the practicalities of form filling for final Disclosure and Barring Service checks so that Sarah would be fully prepared to return in the autumn term to observe the team engaged in the delivery of the prevention programme.

Leaving so soon?

Having traversed chaotic and bumpy ground to reach this point in the study and having shared both hopes and horrors along the way, Sarah was 'knocked sideways' when she experienced a further challenge at this point in her research journey. Having made this initial access and having had such a successful first day of fieldwork, she was not expecting to encounter further roadblocks. But the very day that she was due to step back into the school in the autumn term in 2012, was the day after her partner of twenty-three years had broken the devastating news that he had been diagnosed with stage 4 bowel cancer. The prognosis was bleak. Sarah immediately contacted the school to let them know that her research plans were now delayed, due to this family crisis. She assured the school that she would get back in touch. However, this never happened, leaving Sarah, like Alexandra, with a great sense of guilt and professional distress.

At some point in the future Sarah's research did continue, and she did eventually make it on to an end point where she was able to submit the results of this PhD research for examination. At the time though, her early exit from the school was one which seemed like a 'black hole exit' and which felt like reckless abandonment. Reflecting on these events, Sarah is now able to view this as a form of 'folding in' of the research (because of the complexity and chaos of compounded losses) rather than a final end point which offered no point of return. Therefore, her tale is not entirely one of doom and gloom but, rather, a cautionary tale that research and research endings, like life, can be messy, chaotic and contradictory. Things do not always go to 'plan', instead they can take on a number of unexpected twists, turns and diversions. And sometimes, as Sarah's tale particularly demonstrates, it is exactly because of the messiness of our personal lives that our research takes on these unanticipated and chaotic forms.

Conclusion

These tales have been brought together in this chapter to demonstrate the variety of different, problematic and unanticipated endings which may be experienced by researchers as they leave the field. Indeed, the tales are far from identical. Sarah's exit story was one of an abrupt ending, where her personal life took over her ability to engage in the fieldwork at the time and in the ways originally planned. Alexandra's ending wasn't abrupt, but it was unanticipated and troubling in a variety of ways, even despite the careful planning which had been put in place. Having been opened up for critical scrutiny and shared together in this way, however, the stories do offer some hope for researchers who might find themselves in similar exiting predicaments. And it is in this vein, of sharing for future discussion and learning, that the will chapter conclude. It will do so by reflecting on two collective lessons learned and with a note on how we might move forward in our thinking about exiting, both literally and metaphorically.

The first lesson learned by these researchers was of the necessity of planning. At first sight this may appear perverse, given the unexpected nature of the challenges they were presented with. Fieldwork is necessarily a messy business because of the multiple factors which sit outside of the researcher's control (Watts 2008). Even the best laid plans cannot guarantee a smooth execution and ending to research. But, in looking back on these incidents, both Alexandra and Sarah can recognise the value that their initial research plans held when they attempted to overcome the challenges they faced in the field. In Alexandra's project, the strong participatory principles underpinning the research gave something steady to return to in a moment of panic, and they helped her to make decisions about the best ways to move forward for the benefit of the entire group. In Sarah's case, a steadfast commitment to a feminist ethics of care worked to support her responses to the 'in the moment' ethical disclosures she witnessed. What these researchers learned retrospectively, then, was the importance of building enough space into their research plans: to allow time to react and respond in moments of unanticipated chaos or crisis, but also to 'hold lightly' to these plans and to value the principles underpinning them, as much as the logistical details contained within them, because of the certainty and direction that these could offer even in the midst of uncertainty and wider change.

The second lesson learned chimes with advice given by Michailova et al. (2014) about the positives which can emerge from even the most negative research experiences. For Michailova et al., the positive which emerged from their chaotic exit was a reorientation of their theoretical thinking. These authors remark on the 'revelatory' nature of their exit experience – spurring them on to think about their data in different ways, challenging their existing insights. They contend that acts like this, of discontinuity, can prompt creativity and reflection. The stress and disequilibrium, they say, can spur on new forms of exploration and learning. While the volatility experienced by Sarah and Alexandra didn't necessarily challenge their theorisations of their research foci, they did prompt different ways of thinking about the data which had been generated.

For Alexandra, one positive resulting from her exit experience was that it led to a renewed understanding of the photographs which had been generated in the research and the ways in which they were being used to construct particular gendered subjectivities. For example, as she sat and discussed the images with the parents and teachers of the girls in the meetings which resulted from the bullying incidents, she was initially confused by the conversations. She wondered why the group were debating the potential existence of a nipple in an image which depicted a girl in a low-cut top, when she literally could not make sight of the 'offending item'. But, given time to reflect, she came to realise the importance of this group securing 'evidence' of the girls' sexual innocence in these images – to shore up the 'nice girl' femininities that the school were so keen to construct, and which appeared to offer these girls some safety (at least in the short term in relation to these bullying incidents). This 'jolting' also led to enhanced methodological understanding, as Alexandra came to realise anew the polysemous nature of images and the ways in which they were invested with different meanings at different times as they travelled through various contexts. As she reflected on the changing reception of these images (from 'innocent and girly' to 'dirty and slutty' and back again) she better realised the portable and context-dependent nature of the images and the potential ramifications of this for her research and her participants (Sontag 2003).

In Sarah's case, the positive lesson which emerged from her exit experience was owing to an acknowledgement that life is inextricably bound to research. While this may appear obvious, particularly when undertaking sensitive research, it is a matter which has been somewhat sidelined in the wider research literature. This is particularly the case in relation to those tales which have been told about research endings. For Sarah, this lesson took a long time to be absorbed and acknowledged, most likely because of the lack of discussion on this topic. But eventually it enabled her to reconsider her initial exit experience – no longer seeing it as entirely shameful and embarrassing and resting solely on personal failing and inadequacy, but as a diversion to a different pathway which would lead to an alternative end point for the project. And while the horrific natures of some of these experiences continue to be relived and remembered, Sarah has now reached a stage where she feels reinvigorated in research terms and where her commitment to ethical practice in research has been reinforced. For example, she no longer sees those moments of intense disclosure from participants during her first day in the field as a failure, just because she had to walk away from them at that moment in time. Rather, she regards these more as momentary negotiations which were undertaken at a particular point in the project, and which did reach some resolution, because of her commitment to a feminist ethics of care and the manner in which she was able to stand back, to listen and to put boundaries in place for the participants.

Because of these potentially positive lessons and consequences, and the fact that both Alexandra and Sarah went on to complete their research, some may wish to question the 'black hole' nature of these experiences. While it is the case that some scientists who have researched (literal) black holes do think that they have found evidence of material eventually emerging from them,[1] it is rarely the case that metaphorical black holes are considered in the same way. Scholars more often make use of this metaphor to signal doom and destruction, likening it to a vacuum with no point of return. But neither Alexandra's nor Sarah's research could be considered as having been entirely drawn into a void. The very fact that they remained in academia and continued onwards to publish these tales suggests otherwise. It is also the case that, in recent years, the terminology underpinning the black hole metaphor has been challenged. Some have commented on its racial implications and have questioned whether the language used furthers negative and damaging understandings of blackness (i.e. as unpleasant, destructive or hostile).

It is for these reasons that this chapter concludes by proposing an alternative metaphor for thinking through these 'less than perfect' research exits – that of the labyrinth. The labyrinth is considered to have emerged in Greek legend, as a type of prison-maze holding King Minos' (half man, half bull) Minotaur. It has been used by many scholars in a symbolic sense too. For example, by gender scholars seeking a less rigid metaphor than that of the 'glass ceiling' to explain the multiple barriers women experience in leadership (Eagly and Carli 2007). The labyrinth is commonly taken to symbolise a pathway or journey – one which is meandering (entailing complexities, detours and dead ends) but purposeful (always leading somewhere, and with a pathway which does eventually lead outwards). Because of the hope that the labyrinth metaphor offers for a purposeful pathway through, it may offer a more productive way of thinking through the challenges researchers face in exiting the field. Of course, research projects will sometimes 'fold in' on themselves and researchers might not continue in their endeavours. But more commonly, researchers will find a way through, even if it entails finding an alternative strategy and ending in a way which differs from that anticipated. The labyrinth metaphor not only accounts for this possibility, it also lays down a challenge for us to further explore these multiple, alternative pathways so that we can continue to add to the scrutiny and integrity of future exiting strategies.

Note

1 This is a reference to Stephen Hawking's thesis that when a black hole collapses not all of the information of the star (which had died and formed the black hole in the first place) is lost but, rather, it is encoded in the particles that radiation emits.

References

Bloustien, G. (2003). *Girl Making: A Cross Cultural Ethnography on the Processes of Growing up Female*. Oxford: Bergahn Books.
Delamont, S. (2016). *Fieldwork in Educational Settings: Methods, Pitfalls and Perspectives*. London: Routledge.
Eagly, A. and Carli, L. (2007). *Through the Labyrinth: The truth about How Women become Leaders*. Boston, MA: Harvard Business School Publishing Corporation.
Edmondson, A.C. and McManus, S.E. (2007). 'Methodological fit in management field research'. *Academy of Management Review*. 32: 1155–1179.
Gobo, G. (2011). *Doing Ethnography*. London: SAGE.
Hey, V. (1997). *The Company She Keeps: An Ethnography of Girls' Friendships*, Buckingham: Open University Press.
Lofland, J. and Lofland, L. (1995). *Analyzing Social Settings: A Guide to Qualitative Observation and Analysis*. Belmont: Wadsworth.
Michailova, S., Piekkari, R., Plakoyiannaki, E., Ritvala, T., Mihailova, I. and Salmi, A. (2014). 'Breaking the silence about exiting fieldwork: A relational approach and its implications for theorising'. *Academy of Management Review*. 39(2): 138–161.
Sontag, S. (2003). *Regarding the Pain of Others*. New York: Picador.
Watts, J.H. (2008). 'Emotion, empathy and exit: Reflections on doing ethnographic qualitative research on sensitive topics'. *Medical Sociology Online*. 3(2): 3–14.
Weiss, L. and Fine, M. (2000). *Speed-Bumps: A Student-Friendly Guide to Qualitative Research*. New York: Teachers' College Press.

EXEUNT OMNES!!

The case for bad exits in ethnography

"**Exeunt Omnes**" is a stage direction that translates as "Everybody out!" and it tells everyone to leave the stage hell for leather.

Early in my fieldwork reading I noticed that very few methodological accounts talked about leaving the field — so I imagined that when the study was over someone stood up and shouted "Everybody out!" and that was that. Clearly, I was left with the wrong impression. There's more to it!

What follows here is a different account — a meditation on leaving that explores the idea of the "bad exit" — and as I began with a stage direction, this piece is presented across five Shakespearian acts:

Act One: The Exposition - What is *leaving*?
Act Two: Rising Action - The research context
Act Three: CLIMAX! - The Bad Exit
Act Four: Falling action - What is *leaving*?
Act Five: Denouement!

Act I. An exposition on leaving.

What is leaving?

We leave so much more behind than we know, failing to "leave no trace," (which was never possible) but also taking more with us when we do than just the data. Changing or ending relationships is never easy, even (or especially) for ethnographers.

I set out to understand what it means to leave the field by incorporating ethnographic memoing into CBR freedrawing (Galman, 2021), using Jain's (2019)* generative listmaking technique. I created, then, an analytic list that might "generate queer ways of tracking scents, releasing doves with diverse messages to explore new flightpaths between pigeonholes" (Jain, 2019 p 3).

This CBR technique is also good for flushing out perhaps unconscious, hidden or unseen researcher subjectivities, hot spots, cold spots, and deep grammar. Notably, as I drew I realised that I constructed leaving as mostly liminal ---- and a *difficulty* to be MANAGED Why can't leaving be generative, joyful... a boon? This pricked up my ears as I analysed and told the story of the Bad Exit.

* A fellow anthropologist and artist, Jain reminds us all that, in the words of John Berger, "we who draw do so *not* only to make something visible to others, but also to accompany something invisible to its incalculable destination" (Berger, 2015 p 9, in Jain, 2019)

Exeunt omnes!!

Things That Leaving

Leaving — places, people, workplaces, relationships (and the ethnographic field is arguably all of these things) — is not easy. If it were, there would be fewer country music songs and a lot more ethnographic accounts written about it. I think our assumptions and models are to blame. The "Breakup Model" implies that leaving must be painful, while the "Leaving the Party Model" implies that there is a "good exit" and a "bad exit" reliant totally on the skill of the ethnographer herself. Neither are useful models.

And most ethnographers, so careful with the dance of rapport, practice, and relationship in the field, really want to stick the landing.

I know the Breakup Model. This model comes in both the "Painful/Anguish" and the "Painful/Awkward" varieties. It implies total severance, no contact, no return, and a hard stop. It can hurt.

When I left my horrible marriage it felt like jumping off a moving train in the darkness, black as pitch and cold with rushing terror and the roar of wind and of metal as I stood with my fingers numb from holding onto that cold, sharp edge until

I jumped with my children, the bravest thing I have ever done

52 Entanglements and im/perfect exits

...and I know the Leaving the Party Model, which is absolutely exhausting because it is made of ritualistic etiquette rules*.

Make a good exit!

1. Leave with precision, while the party is still going. Leave before it has a chance to get weird.

2. Don't be the first to leave, but also don't be the last. Be vigilant for useful lulls in the conversation.

3. Say a quick, quiet and crisp goodbye. Make it fast.

4. Don't leave anything behind but if you can't find your coat just smile and pretend you didn't bring one.

5. Be watchful for signs that the host wants people to leave (she is cleaning up, turning on lights, looking at her watch.)

6. Thank the host as you go and send a note of thanks via post the next day. (no email! they might write back!)

In the U.S., this is sometimes called a "Bread and Butter" note.

* these are actual rules my mother had to learn.

Act II Research Context

Beginning in 2015 I have studied transgender and otherwise gender diverse children and youth in families, schools, and communities. As I have written elsewhere (Galman, 2017, 2018a, 2018b, 2018c, 2019, 2020a, 2020b), this work, like much ethnographic work, is an untidy mix of human relationships, carefully built rapport and the many loose threads of the mutinous and messy everyday. It defies our neat procedures and attempts at linearity.

I entered this work as a researcher but also as a mother of a transgender child, and as such very much an insider — what Adler & Adler (1987) might have termed a "complete member researcher" who did not "go native", per se, because I was already there, but for whom political events (Galman, 2017) starkly demonstrated how I could not pretend at scholarly distance. I had not just thrown my lot in with these vulnerable people but instead realised that they were me and I was them all along.

I never lost sight of my research objectives, but in the horrors of Trump's America my participants and I were afraid together for years, and that's a party you can't crisply or quietly or quickly leave.

Entanglements and im/perfect exits

Act IV · Leaving?

A "bad exit" is awkward and uncomfortable and breaks all the rules of decorum in the name of clarity and relationship and putting aside any lingering ideas about being even remotely objective, because you have made yourself vulnerable in the quid pro quo field and that is a feature, not a bug.

"I awkwardly spelled everything out, which would be weird in normal relationships."

[SUPER WEIRD]

"So, hey informants... Study's OVER and I'm going to stop being here all the time as a researcher but I'll still be here sometimes as a friend and I'll always be THAT plus also an advocate but I'll be here less because it would be weird if I was, like ≈poof!≈ gone because what would we tell the kids and also I will specifically say I'm not going to be gathering any more data but if I wanted to in the future but I will be clear about that with new consent forms and stuff so now I want to be here still as a person but not as the researcher to help out and stay connected but you don't have to because the study is over but we are still in this together what do you say? We can be friends but seriously no pressure. I can also just leave. Or not."

What's the word for what we are now?

"Whoa that felt socially strange."

"It might have been easier for her to just sneak out."

"But we are all glad she didn't. This is better."

Exeunt omnes!!

"Also because I am "the researcher", and because of the privileges associated with some of my position(s), I might have had an easier time navigating a transphobic world. I'm done being "the researcher" but I still want to help and use what I have to benefit all of our kids. I am still here.

I delved into the complexity of responsibility and of relationship that does not end or diminish because the research is over.

Sharing all that we did meant I could not just take the data and run like a "thief in the night" (Terkel, 1974, p xvii) but instead use the messy exit process (in which I left but did not leave) "to break down the artificial distance between the researcher and the researched and force us to examine whether our existing methods honor the humanity that we share with our participants"

The work of shepherding relationships through such a transition is also an essential piece of fieldwork, and we must try our best, even if it feels like a "bad exit" to our limited, even positivistic deep grammar that craves the clean, finite, and businesslike end, void of messy humanity.

I still have relationships and responsibilities here.

(Mangual Figueroa, 2014 p143)

Act V: Denouement

"denouement" literally translates to "untying the knot"

When my comic journaling fleshed out my mostly negative or uncomfortable subjectivities around leaving, I pricked up my ears. Leaving the field can be positive if it is done in a way that honors its complexity, accepts awkward conversation and humanizes all parties concerned.

This might also entail rethinking our existing models for leaving.

"Bad exits" can humanize the researcher and the researched, and might helps us in our quest to be "worthy witnesses," more fully human, both entering the field and leaving it.
(Paris & Winn, 2014)

Untangling this knot is too delicate for tidy prescriptives except to encourage EVERY researcher to include LEAVING in their writing → and in doing so, to lay bare the complexity and the awkward humanness of being that "worthy witness." Better to make a "bad exit," a messy exit, a human exit, leaving the dispassionate researcher trope behind and instead allowing/admitting that we, too, have been present, and have been transformed.

References

Adler, P.A. and Adler, P. (1987). *Membership roles in Field Research*. Newbury Park, CA: Sage Publications.
Atkinson, P.A. (1997). *The Clinical Experience*. Aldershot: Ashgate.
Delamont, S. (2002). *Fieldwork in Educational Settings: Methods, Pitfalls and Perspectives*. (2nd edn). New York: Routledge.
Fine, G.A. (1983). *Shared Fantasy*. Chicago: University of Chicago Press.
Galman, S.C. (2017). 'Research in pain'. *Anthropology News*. May/June: 14–17.
Galman, S.C. (2018a). 'Enchanted selves: Transgender children's use of persistent mermaid imagery in self and popular portraiture'. *Shima*. 12(2): 163–180.
Galman, S.C. (2018b). 'This is Vienna: Parents of transgender children from pride to survival in the aftermath of the 2016 election', in C. Kray, H. Mandell and T. Carroll (eds) *Nasty Women and Bad Hombres: Historical Reflections on the 2016 Presidential Election*. pp. 276–290. Rochester, NY: University of Rochester Press.
Galman, S.C. (2018c). 'The story of Peter Both-in-One: Using visual storytelling methods to understand risk and resilience among transgender and gender-nonconforming young children in rural North American contexts', in A. Mandrona and C. Mitchell (eds) *Visual Encounters in the Study of Rural Childhoods*. pp. 161–175. Camden, NJ: Rutgers University Press.
Galman, S.C. (2019). 'Flat caps and dengue fever: A story of ethnographies lost and found India', in S. Delamont and R.J. Smith (eds) *The Lost Ethnographies: Methodological Insights from Projects that Never Were*. pp. 123–140. Bingley: Emerald.
Galman, S.C. (2020a). 'Parenting far from the tree: Supportive parents of young transgender and gender nonconforming children in the United States', in B. Ashdown and A. Faherty (eds) *Parents and Caregivers across Cultures: Positive Development from Infancy through Adulthood*. pp. 141–155. Berlin: Springer.
Galman, S.C. (2020b). 'Ghostly presences out there: Transgender girls and their families in the time of COVID'. *Girlhood Studies*. 13(3): 79–97. https://doi.org/10.3167/ghs.2020.130307.
Galman, S.C. (2021). 'Follow the headlights: On comics based data analysis', in C. Vanover, P. Mihas, and J. Saldaña (eds) *Analyzing and Interpreting Qualitative Data: After the Interview*. pp. 223–231. London: SAGE.
Hilliard, D.C. (1987). 'The Rugby tour', in G.A. Fine (ed.) *Meaningful Play, Playful Meaning*. pp. 173–192. Champaign, IL: Humanities.
Iversen, R.R. (2009). '"Getting out" in ethnography: A seldom told story'. *Qualitative Social Work*. 8(1): 9–26. doi: 10.1177/1473325008100423.
Jain, L. (2019). *Things that Art: A Graphic Menagerie of Enchanting Curiosity*. Toronto: University of Toronto Press.
Mangual Figueroa, A. (2014). 'La carta de responsabilidad: The problem of departure', in D. Paris and M.T. Winn (eds) *Humanizing Research: Decolonizing Qualitative Inquiry*. pp. 129–146. Thousand Oaks, CA: SAGE Publications.

Paris, D. and Winn, M.T. (eds) (2014). *Humanizing Research: Decolonizing Qualitative Inquiry*. Thousand Oaks, CA: SAGE Publications.

Terkel, S. (1974). *Working: People Talk about What They Do All Day and How They Feel about What They Do*. New York, NY: Pantheon Books.

Thompson, K.D. (2018). 'When I was a Swahili woman: The possibilities and perils of "going native" in a culture of secrecy'. *Journal of Contemporary Ethnography*. 48(5): 674–699. doi: 10.1177/0891241618811535.

3

Reflections on care and attachment in the 'departure lounge' of ethnography

Alex McInch and Harry C.R. Bowles

Introduction

'Leaving' is often construed as the end of a period of data collection and a transition point away from the interpersonal dealings of fieldwork to the isolated exercise of writing the research report. This is unsurprising, given that the practice of exiting the field is rarely, if ever, acknowledged in the planning phases of research. Instead, preparatory considerations tend to focus on the initiation of fieldwork and the relational dynamics of gaining entry to a group or organisational setting and establishing rapport. This is also reflected in the corpus of confessional literature documenting the interpersonal complexities of ethnographic inquiry (e.g., Bowles et al. 2021; Burgess 1989; McInch 2020; Parker 1998; Sugden 1997), where the process of leaving has seldom been discussed (Delamont 2012).

Of the few examples that have focused on this under-theorised aspect of fieldwork, Iversen (2009) notes that the leaving practices of ethnographers, in terms of timing and exit strategy, may differ for several interconnected reasons. These include instrumental factors such as grant funding or relational issues connected to the researcher's role and participants' expectations. All will have a bearing on how exiting unfolds and whether the experience is blurry or abrupt for researcher and participants. On the way leaving is often conceptualised, Smith and Atkinson (2017: 647) argue that 'too often, ethnographers conflate leaving a study as physical separation in a spatial sense, with the process of leaving along temporal lines'. For Smith and Atkinson (2017), leaving is never seamless, as field-based attachments are difficult to break. Thus, the process of leaving should be performed with care and attention to the needs and feelings of those involved.

Alongside those contained within this book, this chapter attempts to extend these preliminary insights through two researcher-orientated accounts of leaving from two different research contexts – a school (McInch 2018) and an elite cricket academy (Bowles 2018). Following a brief description of the respective fieldwork sites, two first-person vignettes are presented, offering personalised reflections of the authors' separate leaving experiences. The chapter then concludes by comparing the salient features of the leaving stories and unpacking their implications for understanding the characteristics of diverging ethnographic 'exits'.

The fieldwork sites

Alex's fieldwork unfolded in what was the archetypal working-class secondary school that has become a focal point for researchers in the sociology of education over many decades. Fieldwork lasted for one full academic (and near enough calendar) year between September 2013 and August 2014. The school was part of a federation with a partner school in the same locale, the local education authority having enforced this earlier in the decade because of continuous underperformance. The broad focus of the research sought to investigate working-class orientations to (higher) education. Using Pierre Bourdieu's (1995) logic of practice as the theoretical framework, the exploration sought to understand how young people currently navigate the educational field in the context of their own habitus using their requisite types and levels of capital. The research problem stemmed from the perennial issue of working-class underachievement in UK compulsory education which then translates to working-class under-representation in UK higher education. This is in spite of government intervention through educational reform, mainly through the conduit of compensatory education policies such as free school meals (in compulsory education) and widening-access activity (i.e., means-tested grants) in higher education. The main characters of the fieldwork were of course the pupils, and over thirty interviews eventually took place with Key Stage 4[1] boys and girls. Nevertheless, other key actors became prominent as the research unfolded and the head teacher, several subject lead teachers, teaching assistants and other auxiliary staff also played significant parts in helping Alex to discern how contemporary working-class education is experienced.

Harry's fieldwork, on the other hand, was located in a university-based cricket academy that offered an alternative pathway into professional cricket for aspiring players on the periphery of the professional game. Harry's study focused on players' transitional experiences into (and in most cases away from) professional sport, and the process of occupational identity-exploration related to their lives as aspiring, able-bodied and predominantly White British young men. His fieldwork took place over the course of two seasons including both winter training (late October to early March) and summer (early March to late June) playing months. The fieldwork setting comprised of a head coach (coach), a team manager, an assistant coach, an assortment of medical and sport science personnel and a squad of approximately twenty to twenty-five players per year aged between 18 and 24 years. A variety of training, playing, living and leisure locations acted as primary sites of social interaction that together formed something of a 'captive world' for its participants (Goffman 1961: 15). Indeed, the 'cricket bubble' – a phrase used by players in recognition of their immersive cricketing experiences – exhibited many of the institutional characteristics previously associated with academy sport environments by other ethnographic researchers (see Parker and Manley 2017). These features became most apparent during the months of the competitive season, which bound participants to a social, emotional and material cricket way of life into which Harry became immersed.

Cutting ties

Schools are very peculiar places for fieldworkers. In the UK context there are several procedural hurdles that must be satisfied even before the fieldworker arrives at the school gate. Then, when the metaphorical sleeves get 'rolled up' to commence data collection, a new set of identity dilemmas present themselves in terms of managing/maintaining field relations. I (Alex) must make clear that I am a working-class researcher, and this fieldwork was conducted in a working-class secondary school for one full academic year in an urban conurbation of which I am a native. Throughout my time at Grange Hill (pseudonym), I had forged strong relationships with certain teachers, auxiliary staff and the pupils themselves. As a participant researcher, and in discussion with my gatekeeper, part of the agreement of my gaining unbridled access to the school involved volunteering my services as a teaching assistant, break-time supervisor, exam scribe for pupils with additional learning needs and general dogsbody. This reciprocal arrangement worked well for both the researcher and a school that was experiencing a significant bout of fiscal turbulence.

Aside from a select few, relations with the majority of teaching staff were amicable, at best. I suppose fieldworkers who research school environments are fighting a losing battle around acceptance from the outset, because of the guarded stance teachers adopt towards the idea of yet another external agent being in their company – someone potentially assessing their capability as custodians of knowledge in their respective subject areas. The best example of this was the head of physical education's point-blank refusal of my offer of a helping hand anywhere that he saw fit, even though I have an academic and applied background in sport and physical activity (which he was made aware of). Nevertheless, I quite quickly became friendly with the heads of information technology and history, as well as with the teaching assistants, where I came to spend most of my time in between lessons and, indeed, writing up fieldnotes after the formal school day had finished.

As fieldwork progressed, I attending school became routine and I looked forward to each and every episode. You get to see some inspirational things in schools, but, in the main, schools in disadvantaged areas normally come with an emotional cost. You get to live in the shoes of young people, albeit for a brief snapshot of their day. In the early days of fieldwork, I used to wonder why groups of pupils would arrive at school so early and leave later than they perhaps should. You come to learn of the harrowing domestic circumstances and adversity that they are facing. Like Charlie, a boy in Year 9 whose parents had both, tragically, passed away. He had been 'adopted' by his maternal auntie, who also had young twins, and they were all living in cramped conditions a one-bedroomed flat. Or Latesha, a girl in Year 11 whose single mother was a sex worker, which clearly gives a whole new meaning to working from home, especially for someone who was nearing an important milestone in her young life, that of sitting her GCSE examinations.[2] Then there was Derek, a boy in Year 8 who attended the school's (self-funded) breakfast club every day because that, and his free school meal at lunchtime, were reportedly the only meals that he received on a daily basis. The first thought that came to mind was what about weekends and school holidays? As a working-class researcher researching working-class schooling, I often felt (and still do) guilty that I have managed to escape the fate that for which so many young people from disadvantaged backgrounds are destined. Clearly, my working-class habitus remains the same, but I was able to reconfigure it so that it became compatible with the educational field.

I remember my last encounter at the school like it was yesterday. Ironically, it was not even supposed to be my last encounter, although conducting school-based research inadvertently teaches one to be adaptable, if nothing else. Nevertheless, it was mid-August 2014, and it was the day on which GCSE results were released to pupils. This was a poignant moment for me because, for a lot of these pupils, this marked a day in which a maelstrom of emotions swamped their young minds. For some, their fate had already been sealed; but, for a few, there was trepidation as to what might lie ahead and their next steps (regardless of aspiration and attainment). I wanted to be there for each and every one of them on that day. I had spent hours upon hours in their company in a variety of roles. I had got to know lots of them personally. I had answered their questions, I had supported their learning where I could and there had even been a few laughs along the way. I think they appreciated my continued presence, which was in the main, non-authoritarian.

I arrived early and waited at the edge of the hall as they started to trickle in to receive their examination results. I was careful to stand away from the teachers, so as to not steal their thunder when interacting with their students. There wasn't a full turnout for results, which was expected, but the pupils that I had expected to be there were in attendance. After initial conversations with their teachers, they trickled over to me. I could tell how they had performed simply by their body language as they ambled over. For some, they had reached the end of their formal educational journey (for now, anyway), and they were soon to depart in pursuit of joining the labour market, perhaps heading mainly for the retail and service sectors. For others, they couldn't hide their excitement at performing adequately enough to progress into further education, and they asked me a few tentative questions about what their next steps looked like. I congratulated them and reassured them that investment in their education would enrich their lives like nothing else (like it had done for me). This was easier said than done, though, especially for Wesley, a Year 11 boy who was a registered carer for his disabled mum. His young mind was clearly grappling with an emotional crossroads in trying to reach a decision that was maybe beyond his current intellectual and emotional capacity. He was the top performer in the year group and had aspirations of going to university. This is a seamless transition for many, but not for someone whose household relied on him for care and for money earned from part-time employment as a contribution towards the household costs. It was hard enough for me to try to offer as much encouragement as I could without trampling all over what were important responsibilities and, indeed, working-class values that I am all too familiar with.

The pupils started to disperse, and I shook Wesley's hand and waved to a few others as they made their way out of the school building. I'm not sure if they even realised (or were that bothered) that this was the last time we would ever meet. Diane (assistant head teacher and my gatekeeper) came over and asked me to join the teachers for coffee and biscuits in the staff room, but I could sense that the teachers didn't really want me there, so made up the excuse of a prior engagement and left. We were supposed to keep in touch and meet at the beginning of the next academic year, which was due to start in a couple of weeks, but it never came to fruition. And that is how it played out, an unintentional clean break, never to return. My misjudged grand delusions that (I assumed) I had extensively planned for prior to entering the field were nowhere near realised. No UCAS[3] convention workshops with the Sixth Form, no helping out with school sport fixtures, no maintaining contact with some of the teachers that I had struck up friendships with. The ironic thing is that some of the issues that I had written about extensively while in the field actually contributed to this accelerated parting of ways. In 2016, after a sub-standard ESTYN[4] inspection and a prolonged period of budgetary mismanagement, the school was placed into special measures, faced significant fiscal constraints and underwent a huge restructuring that witnessed a round of both voluntary and compulsory staff redundancies. In the immediate months after fieldwork finished, I often pondered how teachers dealt with the natural cycle of compulsory education. Playing such a pivotal part in young people's lives is a rewarding endeavour, even more so when you come to learn of the adversity that many of them face. I never got to find out if Wesley realised his dream of going to university and becoming the first person in his family to obtain a higher education qualification, and it is more than likely that I never will.

From fieldworker to friend

The process of disengaging from the field is layered with relational and ethical complexities. Like the process of gaining *entrée*, exiting fieldwork is a negotiation over which the researcher has only partial control. During the course of an extended ethnographic stay, fieldwork relationships form in different and uniquely personal ways. Stepping away from these relationships, when the motive for collecting data has passed, is not as straightforward as simply bringing fieldwork to a close. The extent to which fieldworkers are able to find separation is dependent on several factors, such as the locality and proximity of the fieldwork site to a researcher's personal and professional life, as well as the basis upon which fieldwork relationships form and evolve. In some instances, where particular relationships are concerned, disentanglement at the point of exit is better characterised as a matter of degree than a 'clean break' from a personalised attachment built over time.

This assertion is certainly reflective of my (Harry) experience of leaving the field. In specific ways, leaving has been more complex and protracted than the process of gaining access ever was. Central to this is my ongoing relationship with 'Coach' who acted as my primary gatekeeper to a university-based cricket academy for aspiring (male) professional cricketers. My relationship with Coach was primarily constructed around a researcher-participant role formation and related power dynamic that arose from my attempts to get to know him and immerse myself within a professionalised sport environment over which he was principally in charge. Though it was never my intention to become attached to Coach, his influence over the academy's culture made his presence a ubiquitous feature of my fieldwork experience.

Coach was often portrayed as a combative figure and difficult personality. Few who worked with him found him easy to get along with. He had lived a rich but professionally precarious career as a cricketer, umpire and coach.

His prolonged effort to survive within the transient and insecure world of professional sport had left a mark on his character. Coach was proud, self-justifying and ultimately distrusting of people – traits which manifested themselves in an uncompromising temperament and leadership style. Nevertheless, I would learn to like Coach, just as he would learn to like me.

Among the challenges facing my relationship with Coach at the outset of my fieldwork were the fractious working relationships that he had with his players, coaching assistants and university employer. At the time, it seemed Coach was very much a man under surveillance, which my early observational presence at squad training every Wednesday afternoon served only to reinforce. Attempting to establish myself in the capacity of a participant-observer required vigilance to historic tensions and a willingness to operate within a hierarchical and unilateral power structure that Coach was fighting to maintain. Given this backdrop, I had to invest significantly in my relationship with Coach if I was to progress from outsider to insider. Though my primary research interests revolved around his players' experiences as aspiring cricketers, I realised that I had to get close to Coach in order to manoeuvre myself into a position where I could begin to get close to them. The early trips into the field amounted to nothing more than a series of ethnographic visits, where I attempted to stay out of the way as much as possible. However, they set the tone for the rapport I developed with Coach, built on the principle of voluntary subservience.

Underpinning my interactions with Coach was the foundation upon which he constructed his authority. Coach presented himself as an insider to the realities of professional cricket. His insider knowledge was intrinsic to Coach's identity and role, which he used as a discursive device to separate himself from others and all those who attempted to encroach on his territory. At the same time, Coach was a 'cricket man' who liked to talk to 'cricket people'. Recognising and being respectful of our similarities and differences in this regard (I am a competent cricketer who has never played professionally) encouraged Coach to trust me. I would listen attentively to what he had to say about cricket, providing Coach with a captive audience for his biographical musings. Our performances during these regular dialogical encounters generated reciprocity between us. Indeed, our communicative rituals served as a means of self-validation for Coach and an invaluable source of insight for me.

Hours spent in conversation with Coach exposed me to both front and backstage regions of his persona. I learned of his working-class upbringing, his aspirations as a schoolboy, his decision to become a professional cricketer and the path through life that he had subsequently pursued (as much out of necessity as out of choice). Accompanying Coach's narrative was an underlying sense of failure, regret and loss for a playing career that never took off, academic potential he had not explored and romances that had not lasted. Beneath Coach's hard exterior were frailties that the nature of our relationship revealed. Through his solitary self-characterisation, it became apparent to me that, despite his efforts to gain acceptance and recognition within the cricket community, Coach had never really fitted in. It meant that by the time my fieldwork had finished, I had developed empathy towards him that I had not anticipated – empathy that was largely absent from his other working relationships. I understood Coach in ways to which others were not privy. I saw him not as a cricketing tyrant or bully, as his players and coaching assistants may have led me to believe, but as a vulnerable man in his mid- to late sixties whose passion for the game replaced all that was missing from outside of it. Thus, when I learned that Coach had lost his job soon after the completion of my research, I was saddened and concerned for his welfare. Cricket, and his responsibilities as the academy's director, provided Coach with not merely a modest income, but also the means through which he maintained a social network. Though no longer reliant on my relationship with him as I once had been, I felt obliged to stay in touch. What was to come of Coach became an unnerving prospect and one which I felt I could not simply ignore.

In the aftermath of his sacking, maintaining contact with Coach was enabled by our respective involvement in the amateur cricket community around the city in which we both resided. As a participating member of a local cricket club, bumping into Coach during the summer was a reliable feature of my weekend leisure pursuits. His regular appearance at matches, as a spectator and (self-assigned) scout for unearthed cricketing talent, assured that our lives continued to coalesce. My continued access to him reinforced my sense of moral duty that I should, at the very least, retain our cordial association. Our contact during the cricket season would spill over into autumn and winter, where our shared interest and emotional investments in the career progress of Coach's former protégés on the domestic and international cricket stage was a constant source of 'catch-up' conversation and reminiscence, amid the sensory cacophonies of Coach's favoured drinking haunts. The content, context and regularity of our seasonal rendezvous reinforced the sense that Coach valued our relationship as a reminder of what he had once been, and made the notion of terminating it seem neither desirable nor appropriate. Accompanying my sense of loyalty was a feeling of guilt that, despite the assistance Coach had unwittingly offered me at the start of my academic career, neither I nor my research had been able to protect his.

In the years that have followed, Coach and I have continued to talk frequently and meet whenever possible. Unlike the friendly relations I established with many of his players, my relationship with him has escaped a neat conclusion, even though our lives are less entwined than they once were. Extrication is complicated by the fact that Coach has become an increasingly elderly, disabled and estranged man – a reality that sits juxtaposed against my own, and his former players', relative success. Indeed, when I compare my attachment to Coach with the fondness for and interest I still have in the lives of the young men I studied, it is Coach's vulnerability relative to theirs and my own that determines my sense of duty to remain close. Regardless of his motive, whenever Coach calls, I am compelled to offer him my company out of loyalty, care and compassion. As researchers entering field spaces with which we are unfamiliar, we do so unaware of whom we will meet and what we will learn about the people we study. It is therefore impossible to predict how we will respond or attach ourselves to what we find. Connection comes with a cost and a sense of responsibility for the other that can endure beyond the finality and proximity of an ethnographic research project. Fieldwork relationships are not necessarily temporary, even if we intend to treat them as such. In the process of initiating and sustaining my fieldwork relationship with Coach, an unlikely friendship spawned to which I feel bound and forever responsible.

Conclusions

The two vignettes presented in this chapter provide an illustration of the interpersonal and ethical complexities of leaving, as well as the process's unpredictability, for which fieldworkers cannot necessarily prepare in strategic or pre-planned ways. Rather, the way leaving unfolds, as depicted by Alex and Harry's respective stories, is subject to several interlocking factors unique to the relationships between researcher, researched and the studied context (Iversen 2009). Yet, in spite of their inevitable differences, both leaving narratives clearly illustrate the social and emotional investment in fieldwork that lingers well beyond the duration and cessation of ethnographic immersion. As many fieldworkers have acknowledged, ethnographers shape and are shaped by the emotionality of fieldwork, which the process of leaving helps to illuminate (Coffey 1999). Indeed, both Harry and Alex remain affected by their fieldwork experiences and emotionally tied to the lives and life chances of those they studied. Of note here is an ethic of care which emerged over the course of both Alex and Harry's research. Accentuated in their memory and subsequent representations of leaving is a concern for the well-being of others and a want to do what is best, regardless of their capacity to take action. While it may not be practicable or indeed desirable for ethnographers to 'plan' their exits from the field in a conventional research-design sense, Harry and Alex's reflections reveal the need for some prospective consideration on what 'getting out' means in ethnography, so as to assist the emotional preparedness of fieldworkers.

Care as a guiding ethical principle is of course contestable. As Hammersley and Traianou (2014) note, there are significant questions about the role and centrality of care as an ethical prerogative in the context of social research. Indeed, they argue, it is far from clear that in all cases researchers should have an obligation to care for their research participants. More specifically, Hammersley and Traianou (2014) highlight how care in its application takes into consideration the specific context of action. Rather than being universal, the concept of care is particular to perceptions of vulnerability, leading to differential treatment and the categorisation of some people as more vulnerable (and thus worthy of care) than others, which may, in turn, contravene other ethical principles. An ethic of care, therefore, does not provide a simple means of navigating the complex interpersonal dynamics present throughout the stages of fieldwork. The notion of care reflected in Harry and Alex's stories of leaving is that of a compassionate disposition developed in response to their individualised attachments to the situations of others. Here, care is conveyed as a feature of their subjective experience of leaving, not as an essential requirement, which, in both instances, exiting the field exacerbated rather than diminished.

In light of Alex and Harry's diverging experiences, it is worth considering Smith and Atkinson's (2017) position on leaving as a final point of comparison. Smith and Atkinson (2017) take the view that disengagement should be performed gently and reflexively and avoid sudden disassociation along spatial and temporal lines. Indeed, they argue that 'hyper-detachment during the leaving process is morally problematic if not borderline unethical' (Smith and Atkinson 2017: 648). While it is uncontroversial to agree that leaving should be performed in a thoughtful, reflexively aware and emotionally considered way (where possible), it is not necessarily a practical or indeed straightforwardly ethical position to hold. Ethical decisions in qualitative research (e.g., what is appropriate in the context of leaving) are situated in kind (Atkinson 2015; Hammersley and Traianou 2012). What if, for instance, a fieldworker (or participant, for that matter) found him or herself in material and/or psychological danger and needed to get out quickly? The sudden (and painful) termination of Alex's fieldwork is a case in point for the organisational constraints that may also come to pass and prevent further research or relational engagements. Schools, for example, are not places that researchers can freely and simply 'revisit' out of curiosity or care for participants once their formalised research stay is over. Nor, for that matter, are elite sports teams. Leaving, however performed, involves situated judgements, regardless of aspirational or ideological commitments. For Alex, though he wanted to return and had plans to do so, the door was closed on him, which he duly accepted. As for Harry, circumstance enabled continued contact with Coach, resulting in a shift in relational dynamics to which he responded. Though neither can be considered a faultless exit scenario, both need to be understood in their respective interpersonal contexts.

The vignettes presented in this chapter characterise exiting as a relational process and one which should be approached and treated organically and contextually. How researchers arrive at a point of departure, and the subsequent nature of their departing, is influenced by a myriad of relational and contextual circumstances which may strengthen, sever or alter fieldwork attachments in unanticipated ways. Leaving, like gaining access, is another relational act within the ethnographic process where decisions based on competing goals and values may have to be made. It is not easy or ethically suitable to prescribe a course of action other than to advocate deliberation. Whether the decision is to stay in or get out, walk away or return, leave alone or stay in touch, fieldworkers are likely to feel conflicted during the unfurling of an ethnographic exit.

Notes

1 In the UK, Key Stage 4 refers to the last block of learning in compulsory education and pupils aged 14–16.
2 GCSE (General Certificate of Secondary Education) is a series of examinations taken by pupils in their final year of compulsory education in the UK.
3 The University and Colleges Admissions Service is the centralised organisation that administers applications for places at UK universities and higher education colleges.
4 ESTYN is the inspectorate for quality and standards in education and training in Wales, at all levels from nursery to adult, including teacher education and training (www.estyn.gov.wales).

References

Atkinson, P. (2015). *For Ethnography*. London: SAGE Publications Ltd.
Bourdieu, P. (1995). *The Logic of Practice*. Cambridge: Polity Press.
Bowles, H.C.R. (2018). *University Cricket and Emerging Adulthood: 'Days in the Dirt.'* UK: Palgrave Macmillan.
Bowles, H.C.R., Fleming, S. and Parker, A. (2021). 'A confessional representation of ethnographic fieldwork in an academy sport setting'. *Journal of Contemporary Ethnography*. 50(5): 683–715.
Burgess, R.G. (1989). 'Grey areas: Ethical dilemmas in educational ethnography', in R.G. Burgess (ed.) *The Ethics of Educational Research*. pp. 55–71. London: The Falmer Press.

Coffey, A. (1999). *The Ethnographic Self: Fieldwork and the Representation of Identity*. London: SAGE Publications Ltd.

Delamont, S. (2012). '"Traditional" ethnography: Peopled ethnography for luminous description'. In S. Delamont (ed.) *Handbook of Qualitative Research in Education*. pp. 342–353. Cheltenham: Edward Elgar.

Goffman, E. (1961). *Asylums: Essays on the Social Situation of Mental Patients and other Inmates*. Harmondsworth: Penguin Books.

Hammersley, M. and Traianou, A. (2012). *Ethics in Qualitative Research: Controversies and Contexts*. London: SAGE Publications.

Hammersley, M. and Traianou, A. (2014). 'An alternative ethics? Justice and care as guiding principles for qualitative research'. *Sociological Research Online*. 19(3): 24.

Iversen, R.R. (2009). '"Getting out" in ethnography: A seldom-told story'. *Qualitative Social Work*. 8(1): 9–26.

McInch, A. (2018). '"Only Schools and Courses": An Ethnography of Working-Class Schooling in South Wales'. Unpublished PhD thesis. UK: Cardiff Metropolitan University.

McInch, A. (2020). 'The only way is ethics: Methodological considerations for a working-class academic'. *Ethnography and Education*. 14: 1–13.

Parker, A. (1998). 'Staying onside on the inside: Problems and dilemmas in ethnography'. *Sociology Review*. 7(3): 10–13.

Parker, A. and Manley A. (2017). 'Goffman, identity and organisational control: Elite sports academies and social theory'. *Sociology of Sport Journal*. 34(3): 211–222.

Smith, K. and Atkinson, M. (2017). 'Avada kedavra: Disenchantment, empathy, and leaving ethnography'. *Qualitative Research in Sport, Exercise and Health*. 9(5): 636–650.

Sugden, J. (1997). 'Fieldworkers rush in (where theorists fear to tread): The perils of ethnography', in A. Tomlinson and S. Fleming (eds). *Ethics, Sport and Leisure: Crises and Critiques*. pp. 223–244. Oxford: Meyer & Meyer.

4

Unfinished business: a reflection on leaving the field

Gareth M. Thomas

Introduction

In his lecture on fieldwork, Erving Goffman (1989: 132) touches on many aspects of the research process, yet on 'getting out', he identifies that there are 'issues' without saying what these may be. This reflects how accounts of fieldwork access, data collection, establishing and maintaining relationships and the defining role of researcher attributes are staples of textbooks, journal articles and colourful confessional tales, yet stories of leaving the field remain marginal (Delamont 2016). Discussing it is not simply a matter of wrapping up but, rather, involves managing adjustments, expectations and relationships. Moreover, leaving the field is a moment where we 'begin the critical process of reflection, analysis, and contextualization' (Feldman and Mandache 2019: 238). While fieldworkers might temporarily leave the field, I focus upon the moment whereby research ceases, marking a rupture of connection.

In this chapter, I reflect upon my experiences of leaving a post-industrial town after a long period of study. Over fifteen months, I was employed as a researcher on a project[1] in Merthyr Tydfil (Merthyr), a post-industrial ex-mining and steel-making region in South Wales, UK. Once a hub of industrial activity, Merthyr suffered a deep decline owing to deindustrialisation and the closure of industries underpinning the rituals, routines and rhythms of community life. Along with being confronted with a deep political abandonment, Merthyr is often subjected to popular stigmatising narratives that feed into pejorative configurations of the working-class via 'poverty porn' outputs (Elliott et al. 2020; Thomas 2016; Thomas et al. 2018). The project involved working with young people living in Merthyr to explore their perceptions and experiences of living in their community. We collected data using a range of different methods, including semi-structured interviews, observations, video and photo methods, focus groups and soundscapes.

With colleagues, I was embedded in this community for over a year. My employment contract ended in mid-2015 and, as such, I left the research site.

I discuss my disengagement from the field with respect to three matters. First, I trouble the popular narrative of 'leaving' as a voluntary and deliberate decision ordinarily made with reference to data saturation, research fatigue and/or competing personal and professional commitments. I discuss how my abrupt departure from the site prompted, for me, questions and anxieties about researcher responsibilities and commitments. Second, I highlight the affective nature of leaving the field. In conjunction with feminist approaches to research (Huisman 2008; Irwin 2006; Stacey 1988), I argue that extracting myself from the research site led to conflicting feelings of sadness, relief, debt and guilt, revealing the interpersonal intensity of the relationship between researcher and participants. Third, I reflect upon my trepidation about writing once I had left, particularly about how my claims would be received by residents. Together with doing justice to residents' complex worlds over a short period of time, I was bothered by the prospect that my writing might further stigmatise a maligned community, and how my hasty exit possibly weakened the opportunity of resolving any points of contention. Having become a fixture of the scene, I was worried about relationships with the participants post-fieldwork and how they perceived my own interpretation of their worlds, especially given that there were no plans to return. This contribution does not sketch out the reasons *for* leaving the field but, rather, what concerns emerged *because of* my departure. I argue that researchers must treat disengagement as a serious matter of analysis – pragmatically and analytically – that rarely leaves us unaffected.

Leaving the field: unfinished business?

One of the most common suggested reasons for leaving the field is data saturation. Researchers are frequently urged to stop fieldwork when research yields diminishing (analytical) returns; we should admit the incomplete nature of research and halt 'when you have gained an understanding of the setting or slice of social life that you set out to study' (Taylor 1991: 242). Likewise, David Snow (1980: 102) argues that the researcher 'leaves the field when enough data have been collected to sufficiently answer pre-existing or emergent propositions, or to render an accurate description of the world under study'. For Snow, researchers must determine how much data is *enough* and fight impulses to stay owing to a sense of 'unfinished business', especially since researchers – and particularly ethnographers – 'usually collect more data than they can handle' (Snow 1980: 105). Snow's exit from the Nichiren Shoshu Buddhist movement was due to both data saturation and practical constraints (e.g., time pressures, personal role conflict, physical/mental exhaustion). Nonetheless, he struggled to disengage, owing to the attitudes of participants, the intensity of some relationships and a sense of indebtedness to his participants.

Other reasons for leaving include competing personal and professional commitments, participant conflicts and research fatigue and boredom (Delamont 2016; Ortiz 2004; Stebbins 1991). In their study of organ replacement, Renée Fox and Judith Swazey (1992: 10) decided to leave, owing to 'participant-observer-burnout'. They found some of their participants' (health professionals) practices troubling – especially their reluctance to accept the limits of the human condition – and felt that their departure was a statement against the perceived excesses and over-zealousness of organ replacement efforts. In her discussion of the physical burdens of research, Tessa Diphoorn (2013: 214) describes how she became compelled by data collection and only the intervention of a concerned colleague prompted her to leave the field. Others have an easier time of departing, such as when participants live a nomadic lifestyle and expect frequent departures (e.g., Kaplan 1991).

Some researchers recommend a slower process of disengagement, in which they gradually wind down efforts. For instance, Tuula Gordon and Elina Lahlema (2003) propose a solution of not making a full exit so as to ease the transition process (e.g., follow-up interviews). Others, though, suggest an explicit cut. Mette Høybye (2016: 468) suggests that this is particularly important in mediated environments (e.g., online spaces) as, otherwise, researchers can remain 'hooked up' with networks indefinitely, and distance is required in order to produce an analytic interpretation. Similarly, Brent Luvaas (2017: 257) proposes that the significance of the 'extraction event has been substantially reduced' due to the growth of social media, as we retain a capacity to remain in touch. For Luvaas, we 'take the field with us' and 'linger in the ethnographic present indefinitely', to the point that retaining distance from participants remains elusive (Luvaas 2017: 257).

These reflections appear to frame leaving the field as a voluntary, deliberate decision. Excepting a small number of contributions (Snow 1980; Taylor 1991), there is little recognition of *practical concerns* as opposed to *research considerations*. For instance, the researchers in Roberta Iversen's (2009) project varied in their disengagement strategies related to their disciplinary and epistemological orientation, researcher and participant relationships, researcher and participant role perceptions and the pattern of research funding. My own employment contract ended in mid-2015 and I had been offered a job role elsewhere. While I would still be geographically near to the research site, my own roles and responsibilities had shifted. This is an ongoing issue for people working in academia who face precarity and insecurity, are employed on short, fixed-term contracts for project-based research and face explicit encouragement to be mobile. For such researchers, practical concerns likely dictate when a study ends and when writing begins. While research arguably 'suffers when it is rushed or concluded prematurely' (Taylor 1991: 243), the decision to leave can be taken out of the hands of researchers, thereby troubling the common narrative which assumes that leaving the field constitutes an active decision made by the researcher.

My departure from Merthyr felt abrupt and premature. It also ignited worry about responsibilities and commitments to participants. Having spent over a year in the community, I was concerned about how my exit would be perceived by young people and workers at the youth centre who were crucial to my fieldwork. I was unsure what obligations I had to them. They had been so welcoming and benevolent that I felt that exiting would be perceived as impolite and ungrateful, despite their reassurances that this was not the case. Similarly, I was troubled by a lingering thought that participants might feel abandoned. This sentiment is particularly acute in Merthyr. Several youth workers conveyed their dismay at previous researchers parachuting into the community, sucking its lifeblood (as 'data') and departing without a trace and with little reciprocation. Likewise, young people were cautious of 'outsider' representations of *them* and *their place* that may leave them feeling betrayed and bruised. Residents were initially sceptical of outsiders, including those from 'the University' (where I was employed), since they may fuel disparaging representations (I revisit this later). I worried that I would be perceived in a similar manner.

One solution I considered, along with reminding participants throughout the project about its end-date, was to slowly remove myself from the community. Unlike Robert Stebbins (1991: 249), who was often 'free to slip away with little or no need to make a formal exit' in his studies, doing so here would have been odd, if not discourteous, and I felt entering a dialogue of closure was a moral and ethical imperative requiring careful thought. This 'weaning' (Hall 2014: 2187) involved my consistently reminding community members about the project end-date, but noting that they could contact me via e-mail and social media platforms.[2] Once it had ended, I occasionally revisited after my exit for social events and catch-ups, but other professional and personal commitments blunted my capacity to do this as often as I would have liked. My drifting away was likely stimulated by the physical distance between the research site and my new office. Many of the people – young people and workers – had also moved onto new pastures, some voluntary (e.g., employment) and some forced (e.g., closure of institutions due to austerity measures). My perception was that many of those I worked with for the study, particularly the young people, had no interest in the research (or me) following their engagement. In addition, I felt that the research team and I would benefit from having some distance from the field and data (Hall 2014; Ortiz 2004).

In this section, I have attempted to diverge from previous accounts of leaving the field to capture how an exit is not always the choice of a researcher. In many ways, our research is 'always partial and unfulfilled, begging for further inquiry and, at the end, always denied completion, fullness, the whole story' (Fitzpatrick 2019: 167). This perception is particularly acute when research *must* end.

Affective fieldwork

In his research on evictions, Matthew Desmond (2016: 336) identifies how leaving, rather than 'getting in', constitutes the 'harder feat for any fieldwork'. Indeed, various researchers identify how they leave the field with mixed feelings, and perhaps occasionally with some relief too (Hammersley and Atkinson 1983). When researchers withdraw, relationships with participants can be fraught with difficulties, especially if researchers align with feminist approaches to research which value close relations.[3] For Judith Stacey (1988: 26), leaving marks an extra inequality in research relationships, that is, the potential and likelihood of 'desertion by the researcher'. Reflecting on her study in a health clinic in South India, Catherine Riessman (2005: 485) describes feeling 'bad' about leaving because her research team had 'provided a needed service'. There may be a sense of indebtedness too, in which the ethical dilemma 'is not how to respond when asked for help, but how to respond when you are given so much' (Desmond 2016: 336). Researchers may simply just miss being in the setting as well. In *Street Corner Society*, William Foote Whyte (1955: 342) wrote: 'I have moved around many times in my life, and yet I have never felt so much as though I was leaving home.'

Such 'affective engagement' (Høybye 2016: 468) is understandable; long-term fieldwork regularly involves forging deep, sometimes permanent, ties to participants, and this can introduce a complex set of considerations and emotions (Hall 2014; Irwin 2006). As Martyn Hammersley and Paul Atkinson (1983: 122) remind us:

> It can sometimes be strange and disorienting for people in the setting to find that the [researcher] is no longer going to be a part of their everyday world. Informants must adjust to the fact that someone they have come to see as a friend is going to turn back into a stranger.

Relationships can place researchers in 'an awkward or uncomfortable position' when they announce their departure (Taylor 1991: 244). Moreover, ending fieldwork can cause a rupture to the self, 'terminating a big part of [our] identity, both personally and academically' (Delamont 2016: 13). I felt many different emotions just prior to, and after, leaving Merthyr. Initially, I felt indebted to many participants. I believed I was always honest about the realities and possible outcomes of this project. Nonetheless, I worried that 'I was gaining much more from this group than they were going to receive' (Irwin 2006: 162). I could offer only friendship, some small gifts and an aspiration that my work would help others to make sense of their world/s. Moreover, in a setting plagued by political abandonment, I worried that I was also deserting them. I had spent a considerable amount of time and energy cultivating relationships and attempting to understand participants' beliefs, norms, values and concerns, and so to leave felt as if I was, in some ways, walking out on them. This sense of abandonment may be attributed to an anticipated mismatch of expectations between researcher and participants. In her study with Bosnian Muslim refugees in New England, Kimberly Huisman (2008) found that her participants' expectations to stay in contact far exceeded her own. Huisman's extended period of time with this community resulted in greater social obligations and her closeness ignited worries around exploitation and betrayal, especially as some participants found it difficult to trust people following the Bosnian War. While I personally did not encounter a mismatch of expectations regarding keeping in contact, this did not negate a feeling of desertion and guilt. Becoming emotionally involved in the community meant that I may have inadvertently, and insensitively, intruded on participants' lives in the name of research and, in turn, to advance my career, without fully confronting the structures of domination (Ellis 1995; Huisman 2008).

I also experienced a sense of loss and sadness, as I had become 'attached' to participants (Taylor 1991). I had established friendships with several participants, particularly the youth workers who were close in age to me; the same skills, tactics and attributes which make fieldwork progress with relative success may make the process of disengagement more difficult. Charles Gallmeier (1991) experienced feelings of alienation, guilt and sadness after leaving the field, particularly as he had befriended several of the participants (minor-league hockey players). Yet, Gallmeier also described feelings of relief on account of no longer having to navigate players' sexism, his irritation with physical fighting and his annoyance at pranks at his expense. Others have discussed feeling relieved once fieldwork pressures had subsided (Irwin 2006; Ortiz 2004; Reich 2015). Stebbins (1991: 248) describes the joy of finishing fieldwork and how such moments 'number among the happiest of my life', owing to 'now-predictable stages of fatigue and theoretical saturation'. In my study in Merthyr, while I was sad at the prospect of leaving, I did also feel some small relief. My exit would free me of some of the worries outlined above as well as the exhaustion and frustrations which sometimes come with fieldwork. There were occasions where I worked late into the night or where I had spent several hours preparing for research activities with young people, only to turn up at a closed youth centre or a school to discover they had forgotten about our visit. Moreover, the research involved liaising with lots of different organisations and its members as well as other stakeholders (some of whom did not get along). This could be challenging and draining, yet, despite some difficult experiences, the dominant emotion on departing was sadness – reflecting, perhaps, the interpersonal intensity of my interactions during the fieldwork.

Here, I have dissected how fieldwork constitutes an 'affective engagement', a process which rarely leaves researchers unaffected – and, in some cases, unscathed. Reflecting upon this is to confront 'the moral complexities of fieldwork' that are occasionally 'dulled by repeatedly grinding it against the same issues', such as rapport, confidentiality and codes of conduct (Bosk 2001: 205). In the third section, I unpack one final concern on leaving the field: writing about a stigmatised place.

Writing about a stigmatised place

> Leaving the field ... is a ritual of completion, the final stage in a rite of passage, where we reintegrate, with a new status, into the community from which we came. After months, perhaps years, of immersion, we extricate ourselves and retreat to neutral ground. We go home. We gain distance. We reflect on our experiences from a comfortable remove. And then we write. (Luvaas 2017: 257)

Researchers can carry a heavy responsibility; they want to minimise harm, but some people may 'suffer from publication' (Whyte 1955: 342). Publishing work can expose participants to injury and open up feelings of exploitation, betrayal and abandonment by the researcher (Stacey 1988). After publishing a book based upon her research with fishing communities, Carolyn Ellis (1995) encountered considerable anger from participants about it, due to their lack of knowledge about her plans to publish,[4] unflattering language and interpretations, and identifiable pseudonyms. Ellis (1995: 86) was unsettled and horrified by their discontent:

> [I]s doing significant research still more important to me than respecting the lives of the Fishneckers? Does another publication prospect outweigh the possibility of more feelings of deceit and betrayal? Does this article make amends or merely reiterate what bothered Fishneckers in the first place? Will my revelations of personal vulnerability and apology serve as 'corrective facework' or will the remainder of my (and their) lost face (Goffman 1967) embarrass and strain our relationships further? Is there any way to be a

fieldworker and not be a 'fink' (Goffman 1989: 125), 'who by definition is operating against the interests of the observed group' (Fine 1993: 272)?

The topic of betrayal has been explored elsewhere. Snow (1980: 115) claims that participants may feel 'conned, betrayed, or used' as they possibly provide information in the spirit of friendship, not as part of a research dynamic. For Charles Bosk (2001), betrayal can occur twice: once in turning relationships into data, and a second time in researchers retiring to desks to transform experience into text. Like Bosk and Snow, and drawing upon Goffman (1952), I was worried that I might be perceived as a 'con artist' who set up 'marks' and, after gaining confidence and 'taking' them (as well as data), disappeared – but without 'cooling out' the marks. I may have told them exactly what they wanted to hear, however genuine and well-intentioned this was – namely, a promise to tell 'the subjects' world from the subject's point of view' (Bosk 2001: 212).

The primary concern, with respect to worries about betrayal, was around publishing. I was anxious that my claims would possibly intensify the wider stigmatisation of Merthyr. How could I talk about the stigmatisation of Merthyr and its residents without attending to and amplifying the toxic, unfair and abject external portrayal/s of them? This highlights the difficulties of writing about the hold of stigma without tightening its grip. This unease was undoubtedly strengthened by holding positive relationships with participants. Describing her research on policing, Diphoorn (2013) expressed her dislike of many participants and, despite their offering valuable and interesting data, dreaded accompanying them during their work. Unlike Diphoorn, I was extremely fond of the people that I had encountered as part of fieldwork. Our interactions were never limited to what can be loosely described as 'research talk'; we played games, joked around and discussed work, social media, TV programmes, food, family, community gossip, among many other things – so much so that it was easy to 'settle down and forget about being a sociologist' (Goffman 1989: 129).

Nonetheless, the analysis of data occasionally revealed information that would likely reflect unfavourably on some participants – or, more commonly, fuel an already venomous representation of *them* and *their place*.[5] I was also worried that revealing this post-fieldwork would not provide me with an opportunity to cross-check these interpretations with participants. My response was to be honest about my own interpretations and to attempt to provide a fair, accurate portrayal – that is, offering an analytical account as opposed to passing judgement on settings, behaviours or situations. This involved not denying the challenges facing young people and others in Merthyr. In his study with drug dealers, Phillipe Bourgois (2003) did not want to contribute to an inferiorising narrative of poverty, while also not sanitising the suffering and destruction on inner-city streets; to do so, Bourgois argues, would make him complicit with oppression by further exerting power over subjects. In a similar way, I did not want to sanitise accounts, yet I made a concerted effort to ensure that subsequent accounts of my subjects' worlds are not read as a hostile depiction of them and their town. In reality, very few participants requested to see follow-up work (e.g., publications) from the project and, when we did talk post-fieldwork, as I did with several youth workers, we frequently discussed other matters instead (i.e., about our personal lives).

Conclusion

In this chapter, I have sketched out my experiences of, and reflections on, leaving a post-industrial town once fieldwork had concluded. I discussed my disengagement from the field with respect to three matters. First, I disturbed the narrative of leaving as an intentional decision. I identified how my (abrupt) exit prompted questions about researcher responsibilities and commitments, such as how/if to keep in touch with participants. Second, I highlighted the affective nature of leaving the field and my conflicting emotions of sadness, worry, debt, relief and guilt. Third, I discussed my anxieties about publications and how my claims would be received by stigmatised residents. The construction of the researcher self occurs before, during and after fieldwork endeavours – including in our writing – showing how conducting such research is a pragmatic, intellectual and emotional achievement (Coffey 1999). Reflecting upon fieldwork is a process continuing long after leaving the field. It raises questions and concerns about roles, responsibilities and relationships. Unpacking this has analytical potential. It allows us to further understand the world/s that we study, bolsters a commitment to reflexivity and offers an opportunity to reflect back on all aspects of the research process (Delamont 2016). This chapter has been neither exhaustive nor final with respect to thinking through the process of leaving the field. Its aim has been more modest: to stimulate a greater level of reflection and attention to disengagement. My hope is that it is not read as narcissistic and unnecessary navel-gazing but, rather, as contributing to a growing literature acknowledging leaving the field as worthy of further attention. Disengagement processes are not uniform. The strategies that other researchers use to leave, and how they think about and account for this, will vary across contexts and will reflect the values and situations of individual researchers and/or teams (Huisman 2008; Iversen 2009). Yet, offering such tales can offer grist for the mill, leading to fresh insights around managing the challenges of leaving sites. Only by writing accounts of this process can we 'make comparisons, improve our techniques, develop better strategies, and learn from our mistakes' (Gallmeier 1991: 231).

Notes

1 The project ('Mapping, Making, and Mobilising in Merthyr: Using Creative Methods to Engage Change with Young People') was part of the ESRC/AHRC-funded 'Productive Margins: Regulating for Engagement' collective (grant number ES/K002716/1).
2 Jennifer Reich argues that social media affects our capacity to fully leave the field and exposes, if not already, the researcher's *personal* self. Indeed, new media makes it 'possible for participants to potentially embed themselves within the researcher's world' (Reich 2015: 408), which raises questions about whether one can truly 'leave' and/or sever ties when they are always reachable.
3 Katherine Irwin (2006: 160) critiques this stance, namely how becoming close to participants 'can do more harm than good'.
4 A participant told Ellis (1995: 79): 'I thought we was friends, you and me, just talkin'. I didn't think you would put it in no book.'
5 For example, many young people blame the stigmatisation of Merthyr and its residents on others (e.g., drug users). Reading this data in isolation would cast an unflattering light on them. However, as I have shown elsewhere (Thomas 2016), I show that while young people may internalise and further enact place-based stigma, the discourses of vilification circulating about their community were often generated, and maintained, by powerful outside agents (i.e., media rhetoric and political talk). Moreover, many of the young people were keen to counter such pejorative representations.

References

Bosk, C.L. (2001). 'Irony, ethnography, and informed consent', in B.C. Hoffmaster (ed.) *Bioethics in Social Context*. pp. 199–220. Philadelphia: Temple University Press.

Bourgois, P. (2003). *In Search of Respect: Selling Crack in El Barrio*. Cambridge, UK: Cambridge University Press.

Coffey, A. (1999). *The Ethnographic Self: Fieldwork and the Representation of Identity*. London: SAGE.

Delamont, S. (2016). 'Time to kill the witch? Reflections on power relationships when leaving the field', in M.R.M. Ward (ed.) *Gender Identity and Research Relationships (Studies in Qualitative Methodology, Volume 14)*. pp. 3–20. Bingley: Emerald.

Desmond, M. (2016). *Evicted: Poverty and Profit in the American City*. New York: Crown Publishing Group.

Diphoorn, T. (2013). 'The emotionality of participation: Various modes of participation in ethnographic fieldwork on private policing in Durban, South Africa'. *Journal of Contemporary Ethnography.* 42(2): 201–225.

Elliott, E., Thomas, G.M. and Byrne, E. (2020). 'Stigma, class, and "respect": Young people's articulation and management of place in a post-industrial estate in south Wales'. *People, Place and Policy.* 14(2): 157–172.

Ellis, C. (1995). 'Emotional and ethical quagmires in returning to the field'. *Journal of Contemporary Ethnography.* 24(1): 68–98.

Feldman, L.R. and Mandache, L.A. (2019). 'Emotional overlap and the analytic potential of emotions in anthropology'. *Ethnography.* 28(2): 227–244.

Fine, G.A. (1993). 'Ten lies of ethnography'. *Journal of Contemporary Ethnography.* 22: 267–294.

Fitzpatrick, K. (2019). 'The edges and the end: On stopping an ethnographic project, on losing the way', in R.J. Smith and S. Delamont (eds) *The Lost Ethnographies: Methodological Insights from Projects that Never Were (Studies in Qualitative Methodology, Volume 17).* pp. 165–175. Bingley: Emerald.

Fox, R.C. and Swazey, J.P. (1992). 'Leaving the field'. *Hastings Center Report.* 22(5): 9–15.

Gallmeier, C.P. (1991). 'Leaving, revisiting, and staying in touch: Neglected issues in field research', in W.B. Shaffir and R.A. Stebbins (eds) *Experiencing Fieldwork: An Inside View of Qualitative Research.* pp. 224–231. London: SAGE.

Goffman, E. (1952). 'On cooling the mark out: Some aspects of adaptation to failure'. *Psychiatry.* 15(4): 451–463.

Goffman, E. (1967). *Interaction Ritual: Essays on Face-to-face Behavior.* New York: Pantheon Books.

Goffman, E. (1989). 'On fieldwork'. *Journal of Contemporary Ethnography.* 18(2): 123–132.

Gordon, T. and Lahelma, E. (2003). 'From ethnography to life history: Tracing transitions of school students'. *International Journal of Social Research Methodology.* 6(3): 245–254.

Hall, S.M. (2014). 'Ethics of ethnography with families: a geographical perspective'. *Environment and Planning A.* 46(9): 2175–2194.

Hammersley, M. and Atkinson, P. (1983). *Ethnography: Principles in Practice.* London: Tavistock.

Høybye, M.T. (2016). 'Engaging the social texture of internet cancer support groups: Matters of presence and affectivity in ethnographic research on the internet'. *Journal of Contemporary Ethnography.* 45(4): 451–473.

Huisman, K. (2008). '"Does this mean you're not going to come visit me anymore?" An inquiry into an ethics of reciprocity and positionality in feminist ethnographic research'. *Sociological Inquiry.* 78(3): 372–396.

Irwin, K. (2006). 'Into the dark heart of ethnography: The lived ethics and inequality of intimate field relationships'. *Qualitative Sociology.* 29(2): 155–175.

Iversen, R.R. (2009). '"Getting out" in ethnography: A seldom-told story'. *Qualitative Social Work.* 8(1): 9–26.

Kaplan, I.M. (1991). 'Gone fishing, be back later: Ending and resuming research among fishermen', in W.B. Shaffir and R.A. Stebbins (eds) *Experiencing Fieldwork: An Inside View of Qualitative Research*. pp. 232–237. London: SAGE.

Luvaas, B. (2017). 'Unbecoming: The aftereffects of autoethnography'. *Ethnography*. 20(2): 245–262.

Ortiz, S.M. (2004). 'Leaving the private world of wives of professional athletes: A male sociologist's reflections'. *Journal of Contemporary Ethnography*. 33(4): 466–487.

Reich, J.A. (2015). 'Old methods and new technologies: Social media and shifts in power in qualitative research'. *Ethnography*. 16(4): 394–415.

Riessman, C.K. (2005). 'Exporting ethics: A narrative about narrative research in South India'. *Health: An Interdisciplinary Journal for the Social Study of Health, Illness and Medicine*. 9(4): 473–490.

Snow, D. (1980). 'The disengagement process: A neglected problem in participant observation research'. *Qualitative Sociology*. 3: 100–122.

Stacey, J. (1988). 'Can there be a feminist ethnography?' *Women's Studies International Forum*. 11(1): 21–27.

Stebbins, R.A. (1991). 'Do we ever leave the field? Notes on secondary fieldwork involvements', in W.B. Shaffir and R.A. Stebbins (eds) *Experiencing Fieldwork: An Inside View of Qualitative Research*. pp. 248–255. London: SAGE.

Taylor, S.J. (1991). 'Leaving the field: Research, relationships, and responsibilities', in W.B. Shaffir and R.A. Stebbins (eds) *Experiencing Fieldwork: An Inside View of Qualitative Research*. pp. 238–247. London: SAGE.

Thomas, G.M. (2016). '"It's not that bad": Stigma, health, and place in a post-industrial community'. *Health and Place*. 38: 1–7.

Thomas, G.M., Elliott, E., Exley, E., Ivinson, G. and Renold, E. (2018). 'Light, connectivity, and place: Young people living in a post-industrial town'. *Cultural Geographies*. 25(4): 537–551.

Whyte, W.F. (1955). *Street Corner Society: The Social Structure of an Italian Slum*. Chicago: University of Chicago Press.

5

Materia erotica: making-love among glassblowers

Erin O'Connor

Love in the field

In the early 2000s, I conducted four years of fieldwork – an 'apprentice ethnography' – at New York Glass, a glassblowing studio in New York City, where I became a glassblower, albeit a modest one. 'Caught up' in fieldwork, my writing addressed 'where the action is', namely the actual, embodied practice of glassblowing, including becoming both a glassblower and part of the glassblowing social world. One aspect left out, however, was the fact that while in the field, I fell in love with a glassblower and, becoming equally caught up 'in love', became engaged to be married. The question of 'love in the field' is typically answered with a code of research ethics addressing power and difference among researchers and subjects. This is valid. Yet, since my 'exit' from the field – when I stopped taking glassblowing classes, working in the glassblowing studio as a teaching assistant for 'hotshop' classes or as a labourer – as well as the long-ago end of that relationship, I have delved further into the meaning, experience and dynamics of 'body' – material feminism, the environmental humanities and critical indigenous theory, blowing glass once a year with my undergraduate students – leading me to wonder about love anew. In this, and building upon my earlier writing on 'matterly imagination' and 'intra-corporeality', I came to understand 'love' as a dynamic entanglement, unexhausted by human romance (see O'Connor 2007; 2016). In fact, thinking about love turns the ethnographer's analysis away from the human 'field of action', revealing its heteronormative, dualist and anthropomorphic trappings. Thinking about 'love in the field', that is, invites the ethnographer to 'exit' the field altogether, not simply as a moment marking the conclusion of fieldwork but as an onto-epistemological break.

Sarkis

I consider my first meeting with Sarkis, a former track athlete, to have been at the furnace the second night that he substitute-taught Intermediate Glassblowing for Paul. Until that week, we had not had any formal interaction, but I had watched him blow glass. He was one of the best in the hotshop – a 'rock star' – and admired for the way that his practice embodied the challenge of the material with tact, grace, precision and vision. A practical pied piper, the molten glass seemed to flow into the form Sarkis intended of its own accord. It was around Thanksgiving and some students asked that he demonstrate how to make a gourd. 'Ugh', I thought, 'we have Sarkis for a teacher and you want him to blow a gourd?!' I wanted to participate in the beauty that I had watched flow from his limbs, to be caught up in the dance that his movements mapped; to model myself, my practice, after that, which seemed at odds with a kitschy cornucopia object. 'So, you want to see it blown with the optics?' he asked the group, following up on a student request. 'Yeah,' the majority responded in chorus. 'Okay, well then you'll have to give me two tries,' and grabbed his top shirt's hem with crossed arms to lift.

Before I knew that the pre-demonstration huddle had broken, a wave of shyness washed over me, as his shirts' hems rose and exposed a trim and strong torso. He pulled down the undershirt, grayed and threadbare, removed a pipe from the warming rack and walked to the glory hole to heat its tip. His rounding chest and prominent, broad collarbone filled out the undershirt that remained drape-like, and his movements were fluid, one somehow always underway before the last one was completed.

He walked toward the furnace on the other side of the platform, though his teaching assistant, who would typically open the furnace door, had not yet arrived. I slipped away from the students, who were caught up in conversation, and hustled to catch up to Sarkis, laying my hand onto the furnace door handle just as he reached for it himself. 'Ah, you're here, thank you,' he said. 'No problem,' I replied, sliding the door open, facing his profile. The heat and glow of the furnace's ton of molten glass bathed his upper body and he started talking: 'It was weird today,' he began, as he lowered the blowpipe into the furnace, twirling to gather, watching the end of the pipe, 'I went to visit my Grandma.' Anticipation and confusion simultaneously came over me. Grandma? 'And she is like 93,' he continued twirling the pipe, gathering glass on its end.

I liked being alone with him, having this private conversation, which somehow quieted the furnace's roar, while the rest of the class waited on the other side of the platform, out of sight. 'She almost died like ten years ago. She had a stroke, but now, now she and her friends decided that they were going to have a class reunion, so now she is walking around her living room twice. She started with her walker, but now she does it without.' 'That's amazing,' I said, sensing how his movements and manners somehow embraced the proximity required to blow glass, feeling myself partnered in the practice. I looked at his face and could see faint freckles behind the frames of his large, black, rectangular eyeglasses swirled with grey and white that wrapped his temples. 'And her hair is like growing in the reverse direction – it's growing in black now.' Each turn of his tale gathered my attention. I noticed his own black hair, its ruler-straight short strands radiating out like tender quills.

'Ok, open,' he said. The roar of the furnace returned. With a light heave, Sarkis withdrew the gather, undulating, luminous, thick with heat at the end of the pipe, primed to become a gourd. Closing the door, I raised the metal trough of water from the pipe cooler and dragged it over his pipe, back and forth, the cool water showering the hot steel, sizzling, sputtering steam spittle. We didn't talk. He lifted the pipe from the cooler and headed toward the group. The glass waved from the pipe's rotating end: 'I only hope that I can be like that when I'm 93,' he called over his shoulder. 'There is no doubt that you will be,' I called back, eyes brimming. A comment he made to the class the night that he first substitute-taught for Paul floated to mind: 'Yup, I can't talk and blow glass at the same time … (pause) but I can flirt.' I released the trough into the water, heard it 'thunk' on the cooler's metal bottom and walked back to the group in his wake.

Back at the bench, students had gathered around the workbench where Sarkis sat cupping the gather with a wooden block; his workpants were cuffed just above his ankles, exposing worn, black leather boots. He popped up and slid the glass into the glory hole, his body silhouette-like before its coiling fires. Twirling the pipe, taking in the heat, he raised his right shoulder to his turned chin to wipe sweat from his mouth and peered down the pipe into the glory hole to reheat it, mounting the glass's sauntering wag with steady cadenced rotations.

Drawing his right foot behind the left with flamenco-like severity, he nimbly quit the heat, and wheeled the hot glass over his shoulder like a flag girl, parting the sea of students. I did not see a gourd, or any form, in the choreography of heat; the practical charisma of which was dizzying. Though Sarkis had landed the wheel of glass onto an arm of the workbench and was already seating himself in it, my attention, lingering in the glory hole, wasn't quite there yet. My senses were chasing him. His movements, concise and eloquent, gestured like a corporeal calligraphy – a script, which my own body stammered and stuttered to comprehend. 'Can someone blow?' he asked. The student nearest to the blowpipe's opening squatted as Sarkis picked up the folded wet newspaper and, rotating the pipe back and forth across the arms of the workbench, rounded the hot glass in his newspaper-lined hand. 'Blow,' he instructed, his left palm kneading the pipe, while spry fingers ran over silver steel, propelling it back and forth over the workbench's arms. The student puffed up his cheeks and blew, while Sarkis, with cocked wrist and flexed forearm, puckered his lips as if in expectation of a kiss, and shaped the glass into a cone. The meeting of fire and water, paper and glass, centrifugal force and centripetal pressure etched a topography of ash into that day's story of the presidential campaign. It was 2004.

I inhaled the newspaper's smoke as he tossed it onto the workbench and stood, stepping onto a footstool. Catholic smells, the ritual smells of my childhood. He shoved the glass into the optic mold and blew. With the same effortless and engrossing choreography, he returned to the bench and shaped the gourd. Was he performing? Were his movements the flirtation? His twirling? I didn't remember seeing him do it before.

Toward the end of class, my partner suddenly remembered that she had to leave while in the middle of trying to complete another gourd. Sarkis stepped in: 'Erin and I can finish it, don't worry.' She left light-heartedly, though my heart stilled listening to my name be brought into his project. How could my amateur body answer his call? Sarkis took the gaffer seat and looked at me, 'Do we really want to finish this? What are we going to do with it?' The words 'it' and 'finish' felt forced. With my partner, I could follow the 'how to' of blowing the gourd, but Sarkis' presence introduced a challenge of imbuing my hands and body with that lived character of glass that animated his own practice. 'I don't know', I responded, daunted, and continued, 'I don't really understand the gourd. The mould makes it hard to see the form. I can't really follow what I'm doing.'

Though we discussed form and execution, he looked neither at the glass nor at the tools, but directly into my eyes the entire time that I spoke. 'I know, that's why I said that we should do it without the optic mold,' he replied, and breaking his gaze, tapped the pipe, flipped the piece and reattached it from the other direction. Not that I saw each step. I only had heard the thwack and subsequent invitation: 'Why don't you heat it from the shoulder up and open up the neck?' He took the piece to the glory hole and, rotating the pipe with one hand, faced me. 'Uh, sure,' I managed, baffled by his trick, and, stepping into the clearing he had made between his body and the pipe, took over its rotation.

As I stood at the glory hole, taking the heat, Sarkis left to help other students, asking a girl from another group to help me. I finished the heat, returned to the bench, picked up the jacks, began to rotate the pipe back and forth, tooling its opening. Like all my pieces that night, it fell to the ground and shattered. I placed the pipe in the barrel and started to clean up, noticing that I did not care about the lost object, but felt swollen with coursing thoughts, feelings and sensations not readily grasped by any moment of the choreographed production.

Caught up in love's glory

During my fieldwork at New York Glass, love and talk about it abounded – flirtation, sex, flings, relationships, marriage and divorces. Yet, I eschewed the question of 'love in the hotshop', despite 'love' having been addressed by sociological giants like Max Weber, Georg Simmel and Talcott Parsons (for overviews see Rusu 2018 and Featherstone 1998). I recoiled at the thought of subjecting the love I felt – both for the material and practice as well as my eventual lover and friends – to sociological function or otherwise. Moreover, I had tacitly learned all too well about the risk of 'talking about love' without theoretical armature having witnessed the condescension for those who tried to do so; no one asked with June Jordan (2002 [1978]), 'Where is the love?' or evoked the 'power of the erotic' with Audre Lorde (1984). The intersection of gender, sexuality, race and ethnicity in this prejudice is glaring.

In my pursuit of understanding embodied knowledge and the development of proficiency, the work of Pierre Bourdieu loomed large. Broadly speaking, I asked how someone becomes 'caught up in the game' through an analysis of the 'logic of practice', that formation of *habitus* (embodied dispositions, or a lived capacity to do) in a 'relationship of mutual attraction' with the *field*, those object-like social relations such that a person comes to believe in the 'game' (for an overview see Aarseth 2016; Bourdieu and Wacquant 1992: 128; Bourdieu 1999: 512). In my own work on glassblowing, I analysed how the 'dynamic of apprenticeship' forged a visceral comprehension and the development of proficiency (O'Connor 2005). More explicitly addressing 'socialized desire', Loïc Wacquant unpacks the *pugilistic libido* of Chicago's South Side boxers – in the *visceral infatuation* of *passion, of love and suffering* of becoming a boxer toward a 'public, recognized, heroic self … [and] ontological transcendence' (Wacquant 1995: 491–492, 501, 514). In this Bourdieusian tenor, desire is read in a socio-erotics of recognition wherein self-consciousness arises with an abnegation of instinctual desire for connection to an objective world.[1] 'Erin and I can finish it,' aimed the flame, the eros of becoming kindled between Sarkis and myself toward the gourd's production; a certain type of 'relationship', of 'making love' was born.

In Plato's canonical love inquiry in *The Symposium*, Eros/eros, both demi-god and disposition, arcs as intermediary between lack and plenty in accord with his birth from Penia (Poverty/Need) and her lover, Poros (Resource/Path). 'Socialized desire' does not arc unidirectionality from lack to plenty but maintains directionality in 'mutual attraction'; we 'become toward' that which offers itself 'socially for investment' (Bourdieu 1999: 512). By attending to this, the ethnographer follows a 'straight logic' focused 'where the action is' toward the formation of self and world (see Goffman 1967). Action consummates yearning in the glory of achievement and visibility. Askew of this trajectory, Lorde, a Black, lesbian feminist, evokes Eros' more archaic parent – Chaos – from whom/which Eros is as the 'measure between the beginnings of our sense of self and the chaos of our strongest feeling' (Lorde 1984: 89–90). In an affective register, erotic power is the capacity for joy and its sharing – there is no heteronormative 'orientation' embodying a 'straight' directionality by which love moves toward an-other (Ahmed 2006: 68–72). Lorde's love plays out among cisgender women, but its Chaotic origin points toward queer erotic entanglements that call for unpacking 'body'.

A mat(t)er of elemental love

With a modicum of skill, I started working for production teams, typically 'exchanging' my labour for experience. On one such occasion, I 'worked the doors', i.e., adjusted the aperture of the glory hole – that raucous hip-height reheating kiln fed by natural gas and maintained around 1500°F (815°C) – which both retains and releases heat as well as allowing for the entry and exit of the vessel. The work was literally dehydrating, the heat relentlessly extracting sweat from my body. After the day's work I returned home, where, coated in sweat, I immediately undressed – or tried to – in order to take a shower. My clothes stuck, attached by the dried salt of my sweat, visible in cascades of white deposits across my navy shirt. In the shower, under running warm water, I could literally feel the salty crystals wash away and, while the analogy is not exact – glass has a random arrangement of atoms in contrast to salt's crystalline structure – I felt so penetrated by the heat that I felt myself to have become glassy, vitreous, crystalline (see O'Connor 2016). In chemistry and everyday masonry, 'efflorescence' refers to the migration of salt from within a porous material – cement, bricks, stone, etc. – to the surface due to water therein; the water of a saturated brick, that is, literally 'carries' the salt to the brick's surface and evaporates, leaving behind the salt 'deposit'.

In chemistry, glass is not an 'element' in the sense of an anatomically unique substance but, rather, a *mixture* (as distinct from a pure compound), typically forged from silica (sand), soda (sodium chloride) and lime (calcium carbonate), and can include a variety of other minerals. Silica is the 'former' – the primary component – and typically consists of high-quality quartz sand (at least 95 per cent silica dioxide), as this yields the 'purest' glass. Soda (common salt) is the 'flux' – that which lowers the melting point of silica and was traditionally derived from burning marine plants (or 'potash' from wood ash). Lime is the 'stabiliser' – that which prevents glass from crystallising and thereby protects it from corrosion by water (efflorescence). These ingredients are mixed by companies and sold like 'ready-made' cake mix, with one caveat – add fire rather than water. New York Glass used a mix ('batch') called 'Spruce Pine', which is shipped in bags weighing up to one hundred pounds. Was my 'glassy state' not connected to or akin with these salty and crystalline natural resources?

From the fourteenth through the late eighteenth centuries, 'Glass Men' felt themselves to be made entirely of glass – sick with the 'glass delusion', Charles VI perceived his entire body to be 'glass, fused and brittle, fired into a whole that could shatter everywhere' (Harris 2015: 39). Perhaps in writing about 'becoming glass' I, like Charles VI, employed 'matterphor' – 'a tropic–material coil, word and substance together ... agentic and thick' (Cohen and Duckert 2015: 11). But my vitreous intra-corporeality was not the 'pristine realization of self as object, flesh and bone fused into excruciating crystal clarity' of Charles VI – a project of self and world formation vis-à-vis said world and others (Harris 2015: 39). Empedocles wrote of the 'love and strife' of elemental becoming, wherein earth, water, air and fire – the four rhizomatic constituents of the cosmos – become each other in unending flux (Macauley 2010: 110–111). The elemental love of 'becoming glass' is not contained by the stage of human action and realisation, that 'straight logic' of desire in which we 'become toward' what we are not. Elemental love, that is, is neither originated nor exhausted by recognition and its onto-epistemology of achievement and visibility.

In viscous intra-corporeality, I copulated, 'bodies-without-human' co-mingling in salty efflorescence. Far from rendering me blissed-out or 'one with the world', this much queerer love – that overcome and subsumed in the hetero-normative eros of socialised desire – put me in touch with that so often lost in the phenomenal appearance of objects with intention: natural resource extraction. Of salt and crystal, how could I not notice the depths to which the glass was rooted and the extractive industries that afforded 'becoming glassy'? Becoming a glassblower is dependent upon mining, be it for silica or the endless other minerals used to effect glass's material properties, as well as fracking and drilling for gas to 'feed the furnaces'. But, in the straight logic of production, the *mater* – that vital fertility of the living earth – is transformed into mat(t)er.[2] The vessels toward which my salt tended that day had been commissioned; they would be sold and erected in a collection, suspending their intra-corporeal origins in the straight logic of appearance, literally resting atop a pedestal or well-lit shelf, resource extraction hidden in gleaming transparency. My elemental love was thereby stabilised like the glass itself, namely 'set straight' – be it by lime, curriculum, art markets or otherwise – and therein oriented by 'the mat(t)er at hand', quite literally by production. This is an onto-epistemology of relationships *between* 'discrete bodies'.

In the queerer love of vitreous intra-corporeality, we 'exit the field' onto-epistemologically, leaving behind the straight logic of 'becoming towards' and the socio-erotics of recognition 'in relation to' an-other(s). With this break, we can wonder about the intra-twinings, porous wanderings, copious corporealities and promiscuity that we always already are without yearning. We can find the courage to answer glassblowing's challenge to think about love, to write love, to reveal the double entendre of 'making love' in the hotshop. Yes, glassblowers 'have sex' with 'each other' in 'relationships' following a 'straight logic'. But so, too, do glassblowers 'make love', i.e., produce objects in the socio-erotics of recognition, wherein self and world are formed via an-other in a 'straight logic'.[3] Elemental love and its opening onto the *mater* of mat(t)er invites us to leave this 'field', namely that 'clearing of' elemental intra-corporeality and matterly promiscuity.

Sarkis' b(l)oom

On the night of the 2004 election, I interviewed Sarkis at Bo's Bar, tucked into the neighborhood behind New York Glass amid a budding art scene and its politics of urban gentrification. Like the neighborhood, its clientele was a mix of African-Americans born and raised nearby and newcomers – white, Latinx, West Indian and Asian – who occupied the garden- and attic-level apartments of the beautiful, four-storey limestones for the going rate of $600/bedroom. Bo's, like most evenings, was full, with year-round Christmas lights dimly lighting the three rooms. Sarkis was surprised to know I had 'only ordered a ginger ale', ordered a beer and joined me at the table. Rather formally, I reviewed the topic that I had proposed to discuss – learning to blow glass, the body, embodied knowledge – and, recording, dutifully began taking notes, should my recorder fail to distinguish his voice from the low din of hip-hop, house and an occasional Euro-American pop tune from the 1980s.

Sarkis spoke of growing up in Texas as a first-generation Chinese-American, driving his pick-up truck from San Francisco to New York City with his dog to start his art career following the completion of his MFA (Master of Fine Arts) at a large Midwest state school, and finding community among artists in an early downtown Brooklyn live–work loft – a modest seven-storey mercantile building around which skyscrapers now rise. Sarkis, a sculptor specialising in glass, earned his bread and butter like many emerging professionals teaching 'how to' courses at New York Glass:

> 'I won't even say anything. I'll just gather a bunch of glass, sit at the bench, pull it into a rose ... you know, the students are all gathering around at that point and they're like, 'Wow, a rose!' Then I'll take it over to the marver, tap it off, and just stand there watching it. And they're like, 'What?' But I don't look at them. So, I'm sitting [sic] there watching it and, you know, 'Boom!' It starts to crack and shatter and then they're all like, 'What?? Wait! Oh no!!' And I just laugh and look at them.'

I smiled with knitted brow, jotting down a few notes, and continued my interview as he drank, wiping the beer foam from atop his sardonic smile onto the knee of his steel-grey polyester Dickies work pants.

Sarkis' b(l)oom is a *pyromenon* – a 'puzzle of natural philosophy' – that presents the paradox of glass's rigid and molten states: the hot 'interior' wants to expand, while the cool 'exterior' wants to contract, the tension of which causes the explosion.[4] Yet, his b(l)oom houses a puzzle beyond its fiery phenomenal appearance – one that reveals how 'fieldwork' in many ethnographic studies of craft attends to such appearances and human intentionality at the expense of less gleaming enterprises. Uncannily, Luce Irigaray considers the kinship of flower and furnace in her inquiry into 'elemental passions'. Therein, the male gaze consummates the flower 'in bloom' for the sake of that formation of male subjectivity. Severed and arrested from both its cyclical movement – opening with day *and* closing with night – and its subterranean stems, roots and earthen matter. The arrested flower, Irigaray muses is 'no longer embracing. No longer embracing itself. No longer a brazier?' (Irigaray 1992: 32). Of this, she asks 'Is it the splitting of that efflorescence? Mildew and crystal. For instance' (Irigaray 1992: 33).[5]

Consider the brazier of Irigirary's flower and the furnace of Sarkis' rose, a contrast arguably of Hestian hearth and Hephaestian forge. Sarkis' b(l)oom indeed severs 'mildew' from 'crystal', the subterranean from that produced in the power dynamics of natural resource extraction toward the formation of self and world vis-à-vis an-other/an object. Was not all of this 'said' and 'done' and 'visible' the object of my fieldnotes, caught up as I was in the field of human action, in the socio-erotics of recognition? Sarkis' b(l)oom is swept into the dustpan and dumped into the knock-off bin – a metal barrel in which unwanted glass is left to shatter. Because of its 'contamination' with floor dust and debris, it is not remelted, but finds itself in the dumpster, which finds itself in the alley, which a private trash collector picks up to go some 'other' place. Yet, in the moment of Sarkis' b(l)oom – that cracking and shattering – it is only to the lesson in tensility that Sarkis, his students and indeed, my notes attend. Unattended are the b(l)oom's matterly origins, which are not selfsame as its causes. Hereby, Sarkis' *pyromenon* teaches that the glass is a *medium*, namely that which is always 'in reference to' an-other. It is without coincidence that he explains learning to blow glass in terms of a 'relationship':

> I was talking to my students about getting intimate with the glass like a relationship. You have to think about blowing a bubble out too far and it exploding like going to Shakespeare in the Park and that is one of the memories that you share together that becomes the whole that you call your relationship. You go into situations with this person and you deal with the situation and you learn more and more and more. A lot of it is not something that you can write down, it's just something intuitive. It's nothing you even think about until it happens. You have to go through these things. As you do this, you become more intimate with the glass. Once you start figuring it out and experiencing the medium itself, and knowing what it can do, you can do more things. ... The reason why some people are really, really good at it and some people aren't, is because some people really, really understand the medium and some people don't. ... Intimacy is direct experience with the medium. It's not steps, it's not making a cup. It's like dribbling on the floor, getting things too hot, getting things not hot enough, it's understanding the medium, it's never going to go right, you just have to deal.

Direct experience 'with a medium' (by which glassblowers unanimously refer to the glass) is not the same as elemental intra-corporeality. In the relational logic of media, namely as intermediaries 'toward' an-other, we lose the very elemental world of mat(t)er. Flower and brazier open and close, ignite and extinguish, akin to the 'double-ness' of women, themselves the furnace of man's birth, placenta and womb, without the unattached glory promised by Hephaestus' forge and crucible, wherein life is twice-born.[6]

The splitting of the Irigarayan flower's efflorescence into mildew and crystal is an elemental imaginary with which to think the politics of sexual difference (Irigaray 1993). 'Straight love', caught up in phenomenal appearance, aims eros 'toward', circumscribing its power/dynamic to an arena of human action – a field – that 'sets' queer loves – material becomings and entanglements – straight. In such an arena and so aimed, how could I not become 'caught up' in Sarkis' practical charisma 'where the action [was]', wondering how my 'amateur body' could respond to his own, could tend toward and 'be filled' thereby? How could I not desire to 'become a glassblower' amid participating in the fires, smoke, heat, steam, furnace roar, that set the stage of his ascendant work? In the heteronormative frame of the field of becoming by way of acquisition – the development of proficiency – I could 'have' it all.[7]

After our interview at Bo's Bar, Sarkis walked me to a friend's apartment, where fellow graduate students were watching an election in which Republican incumbent, George W. Bush, would emerge victorious over Democrat, John Kerry, by morning. 'Hey,' he called from down the sidewalk. I turned around. 'Can I call you sometime?' he asked. I said, 'Yes.' We were engaged to be married two years later. The tensility of the relationship was apparent to the many who said it was a bad idea from the start. The end of the 'relationship' followed quickly.

Sarkis harboured a handful of close friends and was notoriously difficult – a 'bridge burner' with a personal history littered with misunderstandings, breaks and grudges. Amid this, however, no one disputes his magnetic charisma, and many were wooed, to bed or otherwise, by the transformative power of his practical pied-piping. Known for his prowess, Sarkis was a 'hotshop hero' – that very public masculinity of self, world and feats in the arena of the socio-erotics of recognition. The 'pull' of the b(l)oom promises clarity in crystalline performance, yet this clairvoyance comes only with the enclosure of diathetic elementality and orientation thereby. I cannot help but to see the parallel structure of 'becoming a glassblower' and 'falling in love in the hotshop' as the Romantic tale of *Prometheus Unbound*, namely the straight logic of human achievement.[8] Far too young, Sarkis was diagnosed with cancer and passed away in the 2010s. He was forty-five years old.

Conclusion

Writing about 'love in the field' bears questioning that field; this belongs to a different order than the 'exit', a movement achieved upon an entry. Falling in love with, loving and losing love 'in the field' revealed elemental love as the ground of that field of *socialised desire*, wherein persons abnegate 'instinctual' desire for connection with an objective world and thereby achieve both 'self' and 'world'. This was my 'field of research' – that self and world making via a clearing – an 'alienation from'.

In the US, those 'fields' long ago surveyed and staked, hoed and sewed, by settler-colonialists perpetuate themselves in the topography of the field of social inquiry; the *mater* of *matter* overcome, mastered by logics of 'becomings-like' toward and in approximation of[9] like the Greek *pornê* – the woman/wife turned prostitute having left the patriarch's house, I too have wandered; portable fire, brazier, mobile, unstable and ambiguous, I, like so many women 'out of control' by being 'out of place', have turned *pornê*, recoiling from a logic of masculine domination (Bergren 2008: 249). I offer these queer thoughts in memorandum, askew from those intimate relations of Sarkis' b(l)oom.

Notes

1 This has Hegelian roots. See Kojéve (1969).
2 This is not an eco-feminist essentialising of the earth as woman, but an exploration of the masculine domination as discourse to the exclusion of living vitalities.
3 Non-heterosexual relationships are not immune to heteronormativity (see Hollibaugh and Moraga 1983 among others). I was in a 'heterosexual' relationship with Sarkis as a bi-sexual cisgender woman. See also Bergren (2008) on the bed and the logic of masculine domination in Greek thought.
4 On pyromena, see Harris (2015).
5 On the Irigarayan flower as an image of fluidity, see Canters and Jantzen 2014: 79–81.
6 Regarding the uterine furnace, see Eliade (1971).
7 On habitus and 'having/possession', see Rodrigo (2011).
8 Having been bound in chains by Hephaestus after stealing fire from the gods to give to humankind, Prometheus was unbound by Hercules after 30,000 years. The Romantic interpretation of Prometheus shifted away from Aeschylus' tragedy, Prometheus Bound (479 BC–424 BC), to Percy Bysshe Shelley's heroic account, *Prometheus Unbound* (Shelley 2003).
9 I am indebted to the work of critical Indigenous theorist, Dina Gilio-Whitaker (2019), whose work on Indigenous environmental justice continues to shift my paradigm.

References

Aarseth, H. (2016). 'Eros in the field? Bourdieu's double account of socialized desire'. *The Sociological Review*. 64: 93–109.
Ahmed, S. (2006). *Queer Phenomenology: Orientations, Objects, Others*. Durham and London: Duke University Press.
Bergren, A. (2008). *Weaving Truth: Essays on Language and the Female in Greek Thought*. Hellenic Studies Series 19. Washington, DC: Center for Hellenic Studies.
Bourdieu, P. (1999). 'The contradictions of inheritance', in P. Bourdieu et al., *The Weight of the World: Social Suffering in Contemporary Society*. pp. 507–513. Cambridge: Polity Press.
Bourdieu, P. (2000). *Pascalian Meditations*, trans. R. Nice. Stanford, CA: Stanford University Press.
Bourdieu, P. and Wacquant, L. (1992). *An Invitation to Reflexive Sociology*. Cambridge, UK: Polity Press
Canters, H. and Jantzen, G.M. (2014). *Forever Fluid: A Reading of Luce Irigaray's Elemental Passions*. Manchester: Manchester University Press. ProQuest Ebook Central, https://ebookcentral.proquest.com/lib/marymount/detail.action?docID=5405978 (accessed 3 May 2021).
Cohen, J.J. and Duckert, L. (2015). 'Introduction', in J.J. Cohen and L. Duckert (eds) *Elemental Ecocriticism: Thinking with Earth, Air, Water, and Fire*. pp. 1–26. Minneapolis and London: University of Minnesota Press.

Eliade, M. (1971). *The Forge and the Crucible: The Origins and Structures of Alchemy*. New York and Evanston: Harper Torchbooks.
Featherstone, M. (1998). 'Love and eroticism: An introduction'. *Theory, Culture & Society*. 15(3–4): 1–18.
Gilio-Whitaker, D. (2019). *As Long as Grass Grows: The Indigenous Fight for Environmental Justice, from Colonization to Standing Rock*. Boston, MA: Beacon Press.
Goffman, E. (1967). *Interaction Ritual*. New York: Anchor Books.
Harris, A. (2015). 'Pyromena', in J.J. Cohen and L. Duckert (eds) *Elemental Ecocritism: Thinking with Earth, Air, Water, and Fire*. pp. 27–54 Minneapolis and London: University of Minnesota Press.
Hollibaugh, A. and Moraga, C. (1983). 'What We're Rollin Around in Bed With', in A. Snitow, C. Stansell and S. Thompson (eds) *Powers of Desire: The Politics of Sexuality*. pp. 394–405. New York: Monthly Review Press.
Irigaray, L. (1992 [1982]). *Elemental Passions*. London: Routledge.
Irigaray, L. (1993 [1984]). *An Ethics of Sexual Difference*, trans. C, Burke and G.C. Gill. Cornell, NY: Cornell University Press.
Jordan, J. (2002 [1978]). 'Where is the love?' in *Some of Us Did Not Die*. pp. 268–274. New York: Basic Books.
Kojéve, A. (1969). *Introduction to the Reading of Hegel*. New York: Basic Books.
Lorde, A. (1984). 'The uses of the erotic: The erotic as power', in *Sister Outsider: Essays and Speeches*. pp. 53–59 Berkeley: Crossing Press.
Macauley, D. (2010). *Elemental Philosophy: Earth, Air, Fire, and Water as Environmental Ideas*. Albany, NY: State University of New York Press.
O'Connor, E. (2005). 'Embodied knowledge: The experience of meaning and the struggle towards proficiency in glassblowing'. *Ethnography*. 6(2): 184–206.
O'Connor, E. (2007). 'Hot glass: The calorific imagination of practice in glassblowing', in C. Calhoun and R. Sennett (eds) *Practicing Cultures: Taking Culture Seriously*. pp. London: Routledge.
O'Connor, E. (2016). 'Inter- to intracorporeality: The haptic hotshop heat of a glassblowing studio', in I. Farías and A. Wilkie (eds) *Studio Studies: Operations, Topologies, and Displacements*. pp. 105–119. London: Routledge.
Rodrigo, P. (2011). 'The dynamic of hexis in Aristotle's philosophy'. *The British Society for Phenomenology*. 42(1): 6–17.
Rusu, M.S. (2018). 'Theorising love in sociological thought: Classical contributions to a sociology of love'. *Journal of Classical Sociology*. 18(1): 3–20.
Shelley, P.B. (2003[1820]). 'Prometheus unbound', in Z. Leader and M. O'Neill (eds) *Percy Bysshe Shelly: The Major Works*. pp. 229–313. Oxford: Oxford University Press.
Wacquant, L.J.D. (1995). 'The pugilistic point of view: How boxers think and feel about their trade'. *Theory and Society*. 24(4): 489–535.

Part II

Troubling the field

6

Those who never leave us

Jessica Nina Lester and Allison Daniel Anders

Introduction

In a small city located in southern Appalachia in the United States, Greenland Co-Sponsorship and Refugee Services resettled hundreds of Burundians with refugee status[1] for approximately four years beginning in 2007. Riverhill[2] was a predominantly white, Euro-centric and monolingual place that perpetuated neo-isolationist discrimination (Lester and Anders 2013). A river separated the main street and gentrification from abandoned mills and public housing projects. It was on 'the other' side of the river, in the South Prairie Public Housing Project, that Greenland resettled the majority of the Burundian families.

Most of the Burundians we came to know had been named as refugees thirty-five years before, after the 1972 massacres in Burundi. In what is often named the first genocide of the Great Lakes region in Africa, the Tutsi-dominated government and military murdered over 100,000 Hutus and forced another 150,000 Hutus into exile (Lemarchand 1994). Families fled to the Democratic Republic of Congo, Rwanda and Tanzania. In 2007, new Tanzanian laws restricted access of movement and employment outside refugee camps. In response, the United Nations High Commissioner for Refugees (UNHCR) and the International Organization for Migration (IOM) worked to resettle over 7,400 Burundians in the US and over 1,200 in Australia (US Committee for Refugees and Immigrants 2008; State of Queensland 2011).

In Riverhill from 2008 to 2012, we came to know well the children who were born in refugee camps. In the winter of 2008, a small interdisciplinary research team engaged in a series of focus groups with Burundian men and women in order to begin to understand resettlement. During the same time, an English as a second language (ESL) teacher had begun working with Burundian children in the schools she served and reached out to faculty at a nearby university with a request for help with tutoring. In summer 2008, team members spent time with a number of Burundian children in a self-contained classroom at one of the elementary schools, and in autumn of that year Allison, Jessica and other students began tutoring Burundian adults and children. As the children transitioned into general classrooms the following year, we (Jessica and Allison) began volunteering a couple of mornings a week to help the children with classwork. In the meantime, afternoon tutoring continued with the children in the South Prairie Public Housing Project (Anders and Lester 2011). Much of our work was possible because of our close collaboration with a Burundian translator and community liaison, Rukondo, with whom we met regularly. Indeed, the 'trajectory of fieldwork' was shaped by our relationship with Rukondo (Reeves 2010: 329). She, too, remains intertwined in memories of Riverhill.

While the experiences of Burundian families became the central focus of our work, we engaged in observations and interviews with educators and guidance counsellors at local schools, and interviewed school administrators, staff of the local resettlement affiliate, family co-sponsors, and staff at the health department and area housing programme. Our time with the children and their families became a four-year, community-based ethnographic project. Our lives – as collaborators – became interwoven and the work became and remains a cornerstone of our friendship. The many Burundian children we met and from whom we learned – some of whom now have their own children – remain ever present in our memories. Even today, our conversations with one another always turn to wondering how things are going for Burundian families in Riverhill and when and how we might visit again.

In this chapter, we return to our early encounters and address our arrivals and exits from the fieldwork and the impossibility of fully leaving. To begin, we offer a brief discussion of how we situate our work. Then we describe our arrivals and some memories of our work. Next, we share our departures and all that was difficult and impossible to leave. We point to that which was leavable and conclude by returning to all that keeps us – in memory – committed to the work and the Burundians from and with whom we came to learn.

Getting our bearings

In our ethnographic work, we explicitly leaned on a post-critical orientation (Noblit, Flores and Murillo 2004). Post-critical ethnography reflects commitments to examining power in social contexts and in the everyday practice and representations of ethnography. It requires an acknowledgement that ethnographic work is 'always partial and positional' (Noblit et al. 2004: 22) and emphasises attending to one's positionality and practising reflexivity.

Often, post-critical ethnographers commit to emic perspectives and local knowledges in their study of power and inequities. Post-critical ethnographers ask 'how might we interpret things differently?' and produce multiple representations (Noblit et al. 2004).

In our time in Riverhill, we worked consciously in layers (Bochner 2009) and in relation (Behar 1996). We practised recursive reflexivity diligently (Pillow 2003), aware that as White, English-speaking ethnographers, our White racial privilege, English linguistic privilege, middle-class economic privilege and abled-bodied privilege teach us dominance. Consequently, we worked against ourselves (Noblit 1999) in order that we might understand other perspectives in the world.

Although the ideas that inform post-critical ethnography are straightforward, the work was abundant with contradiction and secrets, and always unfolding at a pace that constantly outdistanced us. A post-critical orientation made us sensitive to representation, and we turned consistently to Malkki's (1996) work with Burundians in the Great Lakes region. Malkki cautioned against a trend in the literature that 'too often' positioned displacement 'as an inner, pathological condition of the displaced' instead of a 'fact about socio-political context' (Malkki 1996: 443). We sought to work against representations that might pathologise Burundian experience (Anders and Lester 2015). We saw that often White educators and medical professionals described Burundian children and their parents as a problem (Anders and Lester 2011; 2018; Lester and Anders 2013; 2014; 2018).

Burundian children – now young adults in their twenties – and their families are the ones who abide across space, place and memory for us. We witnessed pain, loss and grief both in stories of forced exile and in the events that unfolded in Riverhill. We turned to Urrieta's (2003) work on loss, grief and identity and his cautions against analytical tendencies to ignore emotion as a form of knowledge. Frequently, we returned to Noblit's words reminding us that: 'Ethnographic research, beyond anything else, involves committing' (Noblit 1999: 6). Ethnographers commit to people, to practices of understanding that centre people and the experiences of their lives, and to a willingness to learn. Committing includes advocating as well, where the ethnographer creates and engages the participants' worldviews. For Noblit, committing is understood, then, as an active process, and 'committing to people is ultimately what this work is about' (Noblit 1999: 6).

Over time, we recognised another practice of committing in which we engaged. We began to commit to memories of Burundian children and their families (Nora 1989), carrying joy from time spent together, as well as 'dread-full' knowings (Bell 1992) that still haunt us in particular ways. We turn to these memories next, beginning with our collective arrivals to Riverhill.

Arrivals

> Like us, if we knew that things were going to be like this, we would have stopped our journey in [Africa], we would have stopped our trip in [Africa]. But now, we have lost everything that we left behind, and that would be hard for us to go back. And at this point we don't have a choice. They told us: 'You that are going to America, you are going to forget the problems that you had in the past and have a better life. (Burundian father)

For Burundian families, much of what greeted them upon their arrival at Riverhill was unexpected. Many Burundian adults shared that they did not experience a 'better life' in the US – as they were promised and as they had anticipated. Employment and housing instability were real concerns for the families. The meagre assistance that was provided fell short of what was needed to pay for rent and food in the US. Access to learning English, although sought by Burundians, was severely limited. In contrast, tracking adults with refugee status into low-wage labour was consistent. Further, the families faced racist and Nativist discrimination (Anders 2016; Lester and Anders 2013). As one Burundian woman shared, 'You find your mind not in peace'. From schools and school administrators not knowing how to support Burundian children, to healthcare systems not offering the mental health support that many of the children and adults needed, there were systemic gaps that threatened the families (Anders and Lester 2011; Lester and Anders 2014; 2018).

Our own arrivals at Riverhill were interlaced with our privileges. For me (Jessica), my memories of arriving at Riverhill are always produced in relation to our work in the South Prairie Public Housing Project. While I navigated other spaces and interacted with other people in Riverhill, my memories of those Thursday afternoons tutoring and playing with Burundian children are the most present for me and what 'arriving in Riverhill' signifies. It was an arrival in a space I had never navigated previously – despite having lived in a nearby town for two years prior to this. It was on the grassy hills in the housing project where I came to know Riverhill and Burundian families. And it is my conversations with the children that remain with me now. Conversations often moved away from schoolwork, as the children and their parents shared stories about their arrivals in the US, their current situations and told stories of all that they had lost in leaving family behind.

There is no time in Riverhill that is not time with Burundian children and their families. And for me (Allison), there was no work without Jessica. They have become inseparable for me. Seldom was I not an ethnographer in Riverhill. Everything was something to be aware of, always in relation to what we were trying to understand. My arrival in autumn 2007 to work at a university in a nearby city, and departure from that job in 2012, mimicked 'traditional' ethnography. Within weeks of my arrival, I met the ESL teacher who asked for help with the Burundian children the following month.

For both of us (Jessica and Allison), we did not know the work without each other's presence. We were both always part of Thursdays with the children, heart-opening, joyous afternoons always also accompanied by a heart-breaking knowing of systems in the US designed for White, affluent success (Bell 1992). Even today, every phone conversation between us turns back to Burundian children and families. 'Have you heard from Rukondo lately?' or 'Did you see what Spiderman posted …?' The children, now adults, are always with us, and we are always with each other in this work.

What stays: sites of memory and habits of engagement

What stay for us are as sites memory (Nora 1989) and forms of engagement with one another and Burundian community members. What Bell (1992) described as the dread has stayed with us as well. Nora described duty memory as 'the sense of responsibility that weighs upon the individual' (Nora 1989: 16) – an imperative to remember. Interested in sites of memory as symbolic, material and functional, he suggested that it is the historian's intimate relation to their subject that is their means of their understanding. We know the imperative to remember in the gratitude we offer to the children and families who shared parts of their lives with us and in our recognition of the permanence of systemic intersectional violence (Bell 1992; Crenshaw 1991) perpetrated against Burundians in their host country.

Over time, what began as Thursday afternoon tutoring evolved into gatherings across the week where talk was shared about life, culture, loss and joy. Soccer games, community gatherings, weddings, healthcare appointments and school meetings became part of our lives in Riverhill, and all that became difficult to leave behind.

While my (Jessica) Thursday afternoons were meant to start at 3pm and end around 5pm, they typically extended far into the night as I sat on porch stoops and listened to all that they shared. One late night, after returning to my home, I e-mailed Allison to share a conversation I had had with one family – a conversation that captured the kind of meaning-making we collectively embraced:

> They told me about how land is split between males and females in 'my culture' and they said 'I don't like it; I would never do that to my daughter. Would you Jessica?' One of the young boys asked me how many children I want and why. He told me the 'most beautiful' time of the day is when all of his family eats goat together and that he loves all of the children being near

him in the house. He wants to clean the house and his partner will do the cooking, 'not because I think the woman should but because I don't like my cooking. It tastes bad' (one of his brothers laughed so hard and agreed). The eldest brother shared that he is the cook in the house when [their mother] does not cook, apparently. He also wanted to know the differences between 'gay' and 'lesbian'. No wonder I was there so long. We talked around lots of stuff. (Jessica, spring 2011)

As we worked in the community, we were deeply engaged in storying and sense-making, which often remained elusive. We met regularly, sometimes to record our conversations, other times to engage with fieldnotes or strategise a next step, particularly in regard to parent advocacy. Some days we struggled to make anger productive (Lorde 1984); on other days we knotted together unexpected insights so that we would not forget them.

Our practices over time formed our habits of communication with one another, and the habits are linked now to our time together and to memories of the children and their families. Riverhill is both a site of memory and a referent to a way of knowing that we share. We re-engage with memory and these practices formed in Riverhill each time that we write and each time that we talk.

Before shared digital documents were common, we co-constructed a field journal of sorts through e-mails to one another. We each had our own respective set of fieldnotes, but we were not always in the same places at the same time. Our communication generated, too, another layer of dimensionality. Over time, we learned to practise vulnerable writing (Behar 1991; 1996). Many times we worked with only tenuous understandings and would write to know (Goodall 2000). Over the years we cultivated trust in the reception of one another's ideas in what is an always unfolding conversation. For Diversi et al. (2020) the act of writing with someone is a composition 'always in-formation, never still, always moving toward that always new not yet known' (Diversi et al. 2020: 5). Collaborative writing reflects trust, sharing and experiencing a place of belonging for ideas and friendship (Diversi et al. 2020). This shared way of being and meaning-making is an ongoing presence in our lives as ethnographers and friends.

The imaginary place of refuge

We storied trauma and loss, as well as our complicity as White ethnographers always benefiting from the White supremacist ideologies and structures deployed to create what became the US and manifest themselves contemporaneously in myriad reproductions of access and privilege. Systems do not create refuge, and we knew that Burundian children would have to survive White dominance in predominantly White and English-speaking institutions. Greenland, the regional refugee resettlement affiliate, placed the families in public housing projects and enrolled them in schools classified as serving a large proportion of low-income students. Raced Black, and positioned as 'other' in what is a racial and nativist hierarchy in the US, Burundians faced systemic intersectional disenfranchisement and harassment (Crenshaw 1991; Huber 2010). School children, educators, medical professionals, housing administrators, police officers and university officials discriminated against Burundians through xenophobia, racism, Euro-centrism and neo-isolationism. A group of teachers met with their state legislator in the hopes of preventing any more people with refugee status from resettling in the area.

Throughout our time, when asked, we welcomed opportunities to advocate with Burundian parents, even as we recognised that our actions were 'not likely to lead to transcendent change and [might] indeed, despite our best efforts, be of more help to the system we despise than to the victims of that system whom we are trying to help' (Bell 1992: 198–199). We were aware of the intersectional discrimination, limited resources and inconsistent support. In some schools, leadership and teachers used subterfuge to test Burundian children without their parents' knowledge and to track them into special education. Deeply concerned, we met with Rukundo and then with families to respond in what ways we could to their concerns and questions. We developed guides about histories of education in the US and about parental rights, special education and the Individuals with Disabilities Education Act. We hired Rukundo to record the guides in Kirundi for a community forum on education for Burundian families. Rukundo named the project 'Kanguka', which in English means 'Wake up'. The families called it learning 'the big secret' (Lester and Anders 2013). We hoped that parents would understand their legal rights and the larger systems of oppression in the US.

Leaving Riverhill: Jessica's reflections

I (Jessica) said goodbye to Riverhill and the Burundian families in May 2011, moving across the country to a new academic post. The morning I left, I made one final stop at South Prairie Public Housing Project – saying goodbye to the children and sharing when I hoped to return for a quick visit. As I journeyed north, I made another stop in a state where one of the Burundian families had moved. We ate lunch together and talked about their current work conditions, and they asked me questions about my next life steps. We were friends. We knew we would find ways to keep in touch – as friends often do. I remember looking in my rear-view mirror often during that long day of driving, knowing I would be back to visit. Within a few months, I decided to make a 'research trip' back to Riverhill. I met with Allison, touched based with the Burundian children and families and stopped in for a meal with Rukondo in the new town she had moved to. Text messaging and Facebook posts were keeping us connected in ways I had not imagined previously. Three years after I left Riverhill, I got a text message from one of the young Burundian men whom I had come to know. He shared that he was getting married and hoped I would attend. I did, bringing along my new husband. In many ways it was like seeing my old friends. We were the only White people at the wedding, and when the wedding reception began they introduced us as the 'beautiful White family'. About a year later, after not having been back to Riverhill for some time, I became pregnant with my first child. Some local friends held a baby shower for me. Hoping to surprise me, Rukondo drove for fourteen hours to see me at the shower and celebrate my new son. When people at the shower asked, 'how do you know Jessica?' she replied, 'We are good friends and worked closely together in Riverhill.' We are and we did. In fact, just last week, she texted a picture to me of me and Allison sitting next to each other, and left a voicemail that said, 'I was calling to check on you, if you remember me. You're my best girlfriend. Okay, hope you are doing fine. Hope to hear from you. Bye.' So, while I left Riverhill in 2011, so much of it stays with me. So many of the people will never leave; they remain part of my story and their memories, voices, journeys – the things I stood witness to – they never leave. I wish to return often.

After Riverhill: Allison's reflections

In my fieldnotes from 2010, I quoted Behar (1991); she wrote, 'I felt a great need to return, to show people my book and my child' (Behar 1991: 347). We were two years into what would become a four-year project, but Behar's need echoed connections I was feeling. When I left Riverhill in 2012 I already had plans to return, to see the families. I could not stay away. I did not want to stay away. I returned for visits in 2013 and again in 2014. I member-checked prior conversations during these visits, but I went back, because I missed the children and their families, and because I was worried about one family in particular. As I wrote this section I slipped into present tense. I miss them still. I hold a return to be an ever-present possibility.

Four years after my last visit in 2014, I returned with my own child. 'I felt a great need to return, to show people … my child,' Behar wrote. I was aware of wanting to recommit to being with, of a need to keep incorporating into my life new memories with those for whom I deeply cared. 'I miss them all,' I wrote the night before we visited. And indeed, back in Riverhill, I had to pace myself. I wanted to hear all at once from everyone how everyone was. My little one, almost two and a half years at the time, was into everything and still we were welcomed graciously. As I reconnected with family members, I was aware that the youngest among them did not remember me. The afternoons reading books together had disappeared. There were no memories of art projects or playing soccer or with my dog. There was no memory of baking together a first cake, chocolate with strawberry icing like the one in a favourite book we had read. There was no memory of our laughter. My presence by some had been forgotten. 'My vivid memories were not "shared"' (Delamont 2014: 65). Yet new memories were created next to old ones that had been lost. Under the hot, midday sun, I watched my little one toddle happily in circles in the front yard with Burundian teenagers and young adults who were no longer the children I had met on another sunny day ten years before.

Leaving irreconcilable tensions

What was easy to leave and leavable were all the tensions inside the university around the work. We experienced tensions at many levels, some more productive than others. The university's commodification of our work was as reductionist as it was celebratory. Multiple university communication teams sold a narrative of a successful university 'intervention' in the problematic lives of 'refugees' for awards and status. The neoliberal success story stood in contrast to what we had witnessed every week in our work (Anders and Lester 2011). We resisted the commodification generally and the narrative specifically, but not everyone did. The narrative splintered our small interdisciplinary team. Some welcomed the attention and press, perhaps thinking they could control the narrative.

Intellectually, we understood ourselves as labourers subject to a system of neoliberal capitalism. We knew the university owned our labour and its exchange value. What we did not know was how to navigate the politics. What we did not anticipate was the spectacle that the university could create of the work. Interviews with team members about our work were packaged as university 'success' for consumption on the JumboTron (i.e., video screen) in the football stadium on game days. A hundred thousand potential donors consumed the message of the university as a 'white saviour' of 'refugees'. The spectacle exploited the violence and forced displacement of Burundians who had endured over thirty-five years in refugee camps and rendered their terror and the egregious conditions of resettlement in the US not just invisible but non-existent.

The narrative that the university sold was the same narrative that we attempted to work against. We were aware of our White supremacist ideological inheritance in what became the US and of the ways it manifests in US public education (Bell 2004). White settler colonialism informed the ideas in the university's narrative and had produced the historic imperial conflicts in Ruanda-Urundi (Rwanda-Burundi). As White scholars we never escape the potential charge of 'white saviour', but we worked to invest our time where Burundians wanted us to spend it.

Responding to a staff member in university communications who had drafted a story about Burundians in the area, Allison attempted to complicate the 'success' story that the university was trying to sell.

> [Burundians] live in public housing projects in [Riverhill] where they are not free from struggle or violence. Many work as custodians on our campus making far below what is a living wage in this country. The result of [Greenland] being unable to find enough sponsors in the area in time for arrivals is important, and real, and has unintended consequences ... the tenor of 'success' is a misrepresentation. (Allison, fieldnotes, winter 2010)

We resisted the pursuit of any awards that might jeopardise the anonymity of the children and families. This was not a concern of university leadership, whose priority was to expand the university's capital. We stopped giving interviews and prohibited photography by anyone who was not Burundian at community events. The spectacle that the university created became one more thing from which we could not protect Burundian children and families in the US. The experiences with university communications made us wary of communication requests we have received since then – even as we work at other institutions. Our respective departures from the institution were exponentially easier than those from the children and families we had come to know.

Conclusions: living, leaving, committing

There were tensions which we easily left when we exited the field; yet, so much remains with us. For us, this work, centred on our commitment that 'when you understand people, when you have committed to them, and when you have learned from them, you advocate for them ... where advocating means trying to promote their world view as reasonable' (Noblit 1999: 8). Memory is a way of knowing (Behar 1991), and duty memory a form of committing (Nora 1989; Noblit 1999). Riverhill remains meaningful work, painful work and work that teaches us again and again. Its ongoing life in our lives always rests in the landscapes of our memories (Delamont 2014), but not only in the landscapes of our memories. Many, though certainly not all, of the friends we made reach out with updates and news to share. These are memories that never leave us.

Notes

1 We use 'refugee status' rather than 'refugee' because Burundians we knew predominantly self-identified as 'Burundian' and 'African'.
2 Pseudonym. All proper nouns and Burundian names are pseudonyms.

References

Anders, A.D. (2016). '"It's almost like we were sold": Burundians with refugee status and educational and economic inequity in the US', in G.W. Noblit and W. Pink (eds) *Education, Equity, Economy: Crafting a New Intersection*. pp. 97–116. New York, NY: Springer.

Anders, A. and Lester, J. (2011). 'Living in Riverhill: A postcritical challenge to the production of a neoliberal success story', in B. Portofino and H. Hickman (eds) *Critical Service Learning as Revolutionary Pedagogy: A Project of Student Agency in Action*. pp. 223–249. Charlotte, NC: Information Age Publishing.

Anders, A.D. and Lester, J.N. (2015). 'Lessons from interdisciplinary qualitative research: Learning to work against a single story'. *Qualitative Research*. 15(6): 738–754.

Anders, A.D. and Lester, J.N. (2018). 'Examining loss: Postcritical ethnography and what could be otherwise'. *Qualitative Inquiry*. 1–11. doi: 10.1177/1077800418784327

Behar, R. (1991). 'Death and memory: From Santa María del Monte to Miami Beach'. *Cultural Anthropology*. 6(3): 346–384.

Behar, R. (1996). *The Vulnerable Observer: Anthropology that Breaks Your Heart*. Boston, MA: Beacon.

Bell, D. (1992). *Face at the Bottom of the Well: The Permanence of Racism*. New York, NY: Basic Books.

Bell, D. (2004). *Silent Covenants: Brown v. Board of Education and the Unfulfilled Hopes for Racial Reform*. New York: Oxford University Press.

Bochner, A. (2009). 'Warm ideas and chilling consequences'. *International Review of Qualitative Research*. 2(3): 357–370.

Crenshaw, K. (1991). 'Mapping the margins: Intersectionality, identity politics, and violence against women of color'. *Stanford Law Review*. 43: 1241–1299.

Delamont, S. (2014). *Key Themes in the Ethnography of Education: Achievements and Agendas*. Los Angeles, CA: Sage.

Diversi, M., Gale, K., Moreira, C. and Wyatt, J. (2020). 'Writing *with*: Collaborative writing as hope and resistance'. *International Review of Qualitative Research*. 14(2). doi: 10.1177/1940844720978761.

Goodall, Jr., H.L. (2000). *Writing the new ethnography*. Walnut Creek, CA: Altamira.

Huber, L.P. (2010). 'Using Latina/o critical race theory (LatCrit) and racist nativism to explore intersectionality in the educational experiences of undocumented Chicana college students'. *Educational Foundations*. 24(1–2): 77–96.

Lemarchand, R. (1994). *Burundi: Ethnic Conflict and Genocide*. Cambridge, MA: Woodrow Wilson Center Press and Cambridge University Press.

Lester, J.N. and Anders, A.D. (2013). 'Burundi refugee students in rural southern Appalachia: On the fast track to special education', in J. Hall (ed.) *Children's Human Rights and PUBLIC SCHOOLING in the United States*. pp. 71–87. Boston, MA: Sense Publishers.

Lester, J.N. and Anders, A.D. (2014). 'Complicating translation: Children with refugee status and special education testing'. *NYS TESOL*. 1(2): 25–38.

Lester, J.N. and Anders, A.D. (2018). 'Engaging ethics in postcritical ethnography: Troubling transparency, trustworthiness, and advocacy'. *Forum Qualitative Sozialforschung (FQS)*. 19(3) Art. 4, http://dx.doi.org/10.17169/fqs-19.3.3060.

Lorde, A. (1984). *Sister Outsider: Essays and Speeches*. Berkley, CA: Crossing Press.

Malkki, L.H. (1996). 'Speechless emissaries: Refugees, humanitarianism, and dehistoricization'. *Cultural Anthropology*. 11(3): 377–404.

Noblit, G.W. (1999). *Particularities: Collected Essays on Ethnography and Education*. New York: Peter Lang.

Noblit, G.W., Flores, S. and Murillo, E.G. (2004). 'Postcritical ethnography: An introduction. In G.W. Noblit, S.Y. Flores and E.G. Murillo, Jr. (eds) *Postcritical Ethnography: Reinscribing Critique*. pp. 1–52. Cresskill, NJ: Hampton.

Nora, P. (1989). 'Between memory and history: Les lieux de memoire'. *Representations*. 26: 7–24.

Pillow, W.S. (2003). 'Confession, catharsis, or cure? Rethinking the uses of reflexivity as methodological power in qualitative research'. *Qualitative Studies in Education*. 16(2): 175–196.

Reeves, C.L. (2010). 'A difficult negotiation: Fieldwork relations with gatekeepers'. *Qualitative Research*. 10(3): 315–331.

State of Queensland (Queensland Health). (2011). 'Burundian Australians. Community profiles for health care providers.' Queensland Health and Multicultural Services. www.health.qld.gov.au/__data/assets/pdf_file/0026/157508/burundian2011.pdf.

Urrieta, L. Jr. (2003). 'Las identidades Tambien lloran, Identities also cry: Exploring the human side of Indigenous Latina/o identities'. *Educational Studies*. 34(2): 147–168.

US Committee for Refugees and Immigrants. (USCRI). (2008). 'World Refugee Survey 2008: Tanzania'. www.refworld.org/docid/485f50d5c.html.

7

Déjà vu et jamais vu: what happens when the field expands in ways that mean there is no exit?

Dawn Mannay

Introduction

'Leaving the field' has been defined as 'the social process of withdrawal from fieldwork' (Bloor and Wood 2006: 112). The categorisation of fieldwork, rather than the overarching umbrella term of qualitative research, is most closely aligned with the ethnographic tradition. Ethnography involves social exploration and protracted investigation, and it is generally relatively small in scale so as to engender in-depth study. Accordingly, for Madden (2019: n.p.), it is ethnographers and ethnography that turn 'someone's everyday place into a thing called a "field"'.

The field encompasses geographical and social landscapes inhabited by people who become participants, and immersion in the field engenders opportunities to draw novel connections and synthesise insights in relation to the local and global context (Fine and Deegan 1996). The toolkit of the ethnographer can include participant observation, relatively open-ended interviews with participants and the analysis of artefacts and documents associated with their lives (Hammersley and Atkinson 2007: 3).

This doing of ethnography is widely discussed and examples, reflections and advice on how ethnographers should enter the field and conduct themselves once in the field form the basis of many handbooks for students and more experienced researchers (Coffey 2018; Delamont and Atkinson 2021; Gobo 2008a). However, although some of these texts offer accounts of disengagement, overall, far fewer books and papers give this stage of the process considered attention (Pole and Hillyard 2017), and 'leaving the field' has been cited as one of the 'most neglected problems in the literature' (Gobo 2008b: 303) and the 'least discussed aspect of ethnography' (Delamont and Atkinson 2021: 137).

This may be in part because leaving the field is not always an unambiguous process. While some researchers are forced out of the field when their work is prematurely curtailed or lost before it begins (see Smith and Delamont

2019), for others leaving is more complex. For example, in her study of racialised misrecognition, Annabel Wilson (2021: 72) discussed an 'ethic of friendship' where, after fieldwork was complete, she continued to meet with participants, abandoning her position as researcher because the conditions of these relationships had changed, and families previously situated as participants went on to be guests at her wedding and her daughter's christening. Similarly, Michael Ward (2015) reflected on returning to the research site of his work with young men in Wales six months after leaving the field, to attend the funeral of nineteen-year-old 'Davies'. Ward described returning both to pay his respects and to support the participants from his research who were attending. Nonetheless, in the space of the church he was conscious of whether he should display the emotions that he felt through crying, because, as a researcher, 'even after knowing them for such a long time, I am still an outsider' (Ward 2015: xiii).

For others, rather than still being on the periphery as a relative outsider there are issues in being embedded in a way that means there is no available exit by which to leave the field. Accordingly, research as an insider within one's own field can problematise ideas of departure. Undertaking a doctoral study in Wales's literary sphere, as an academic, writer and critic, Lisa Sheppard was navigating a field that she would remain within well beyond the doctoral journey. Sheppard (2018: 204) notes the ways in which it was 'emotionally tricky to navigate relationships between authors, critics and lay readers', publishing her work, disagreeing with respected colleagues and seeing them at conferences. These are trials to be negotiated in everyday academic life, but when your research field is also your academic livelihood and home, matters become more complex.

Fieldwork is characterised by 'the messiness of layered subjectivities and multi-dimensional relations in particular localities' (Hopkins and Noble 2009: 815); as such, emotion is a constitutive element the research process (Loughran and Mannay 2018). It is also a site of insider and outsider debates, issues of fighting familiarity (Delamont and Atkinson 1995), and reflexive practice to understand individual subjectivity, while at the same time an endeavour of seeking to understand what it means to be someone else. Reflecting on their fieldwork participants, Lisiak and Krzyżowski (2018: 37) comment: 'we were familiar with them, but it does not mean we knew them'. This knowing and not knowing also extends to how the researcher recognises themselves, and the concepts of 'déjà vu' and 'jamais vu' can be useful in unpicking these shifts and complexities when we consider leaving or, in this case, not leaving the field. Before considering the potential of these conceptual frames it is important to introduce the specific field of focus.

The field

While for Bourdieu (1990) the term 'field' refers to the different dimensions of a society, ethnographic fields are created as a consequence of particular research projects (Madden 2019). The projects discussed in this chapter were very much located in the field of education, and interested in its logic, practices and inequalities, but the ethnographic field was narrower, with a specific focus on the geographical space of Wales and everyday experiences of care-experienced children and young people in schooled, recreational and home spaces.

There are critiques that 'too much research uses interviews rather than ethnography' (Delamont 2012: 57), and being in the field with opportunities to watch, listen and experience is often seen as the gold standard of ethnographic research (Atkinson and Coffey 2002). However, some fields are impervious to forms of direct observation (Lincoln 2012), and in these cases conventional ethnographic techniques may be impossible, risky or dangerous (Coffey 2018). The initial study that brought me to the field was one that was interested in the education of sixty-seven care-experienced children and young people aged from 6 to 26 resident in Wales (Mannay et al. 2015). It would not have been appropriate to follow these participants in school or university settings, where the label of being 'looked after' invariably demarcated them as being different (Mannay et al. 2017).

Children and young people have expressed frustration at being viewed and understood through the lens of being 'looked after' in the care system (Hallett 2015). Accordingly, rather than going into schools, we sought opportunities outside of the school to gain an understanding of the participants' educational experiences in safe spaces where all children and young people were care experienced. It was also important to move away from the format of the social work encounter, differentiating from the forms of communication and interaction that are used to inform care status and decide how children and young people are 'looked after'.

As Coffey (2018) contends, it is possible to conduct high-quality ethnographic research without immersion in a field of study, and situating alternative techniques within an ethnographic framework can offer up 'the potential for engendering a more nuanced understanding of social worlds' (Mannay and Morgan 2015: 170). The initial study (Mannay et al. 2015) incorporated visual and creative activities, focus groups and individual interviews, and care-experienced peer researchers were directly involved in feeding into the design and facilitating group interviews with the older participants (aged 16–27). The all-day events with younger participants in primary school (aged 6–11) and in secondary school (aged 11–16) offered activities including clay modelling; wall climbing, sport-based games; and jewellery, T-shirt and bag making, where participants could take home the items they created. The research activities were embedded in these events but they were optional, and some children and young people joined the days but chose not to take part in interviews or creative tasks linked to the study. This element of choice was important and the sixty-seven participants who did discuss their experiences generated a nuanced data set and shared important messages for policy and practice. This initial project came to a close, and this meant a nominal exit from the research, but not a complete departure.

Not leaving the field

As Pole and Hillyard (2017: 108) contend, the 'field remains part of the repository of experience upon which the researcher may draw throughout' their career, and there was the immediate sense of being in the data and disseminating the data – but this was followed by a more protracted engagement. Initially, the project generated the expected outputs of reports (Mannay et al. 2015), articles (Mannay et al. 2017; 2019) and book chapters (Mannay, Rees and Roberts 2019); but while these outputs may accrue benefits for the researchers, their audiences are often restricted to narrow fields within academia (Barnes et al. 2003; Timmins 2015).

In completing this project, it had been made evident that leaving the field to pursue academic publications did not sit comfortably with me as a researcher. The conceptualisation of the researcher leaving the field, that 'once they have filled their bags, they escape with the loot, never to be seen again' (Gobo 2008b: 306), was familiar to participants in the study. For example, preceding one group interview a young man told me, 'I didn't come for you, you come and ask questions and go, nothing ever changes, I only came to meet people like me who live locally.'

Given the many previous studies in this area and the ongoing inequalities faced by care-experienced young people, which I will return to in the following section, it was not surprising that the young man positioned me in this way. Using the analogy of researcher as a vampire, Ward (2015: 170) described how he 'felt very much like a vampire, trying to suck information instead of blood from my participants or victims' – and I could imagine myself in the black cloak, with fangs, in this moment of recognition. In response to the 'what will happen after' question, I did not want to explain that I would publish but that this dissemination strategy would have little impact on practice, policy or communities, limiting any potential opportunities for change and improvement.

It was important that the project did no harm to the research subjects (Bloor and Wood 2006), but it was equally important to consider moving beyond this avoidance position to explore ways to advocate for the rights and interests of participants (Hugman et al. 2011). The goals then changed to not only fostering ways to listen to the voices of participants as a form of social research but to 'raise the volume of the voices of care experienced children and young people, and to develop mechanisms which can inform, and potentially improve, public services in the fields of social care and education' (Mannay et al. 2019: 59).

The academic outputs generated offered a useful springboard to begin this work, as they informed policy initiatives within Wales (Welsh Government 2016; 2017). However, policy change does not always make changes on the ground, as Sara Delamont discovered when studying teachers who had not 'heard of the legislation' (Delamont and Atkinson 2021: 151). The published works also secured entrance to and funding from opportunities made available by the impact agenda. There has been an emphasis on the potential for research to bring 'change or benefit to the economy, society, culture, public policy or services, health, the environment or quality of life, beyond academia' (REF 2011). This can be viewed as an additional pressure on academics to 'influence the world' and be measured on their world-changing abilities, alongside research, writing and teaching responsibilities (Knowles and Burrows 2014: 242). Yet, the impact agenda can also offer opportunities to remain in the field in a different way.

At the time of writing, I am seven years in the same 'field' of work. There have been other projects with different topics, but a significant amount of my time has been spent working in relation to care-experienced communities. Impact has been key, with multimodal outputs including magazines, music videos, animated films, dedicated websites and artwork; training events and workshops have been held across the UK and Ireland with foster carers, teachers, young people, social workers and related practitioners (Mannay et al. 2019; Kara et al. 2021). There have also been further research projects with different funding bodies and partners, but often this work includes partners from the initial project, and we remain connected in research, projects and advocacy and impact activities. I work with care-experienced young people in different ways, and the same young people have met me in multiple contexts: as the social researcher, as the trainer in research methods for peer researchers, as the collaborator in an animated film project using fuzzy-felt storyboards, as the colleague on a project where a young person has gained a paid role as a research assistant, as a source of advice for study pathways and careers.

These activities contributed to a Research Excellence Framework[1] (REF) Impact Case Study, but there was no discrete linear journey from research in the field to dissemination activities, neither was this extended engagement with the wider field motivated by gaining measurable impact; rather, measurable impact was an artefact of not exiting. Engagement with the field and the actors within it has been an attempt to maintain ethical responsibilities in designing, conducting and getting the messages out from research. The projects and impact ventures are not research, yet they have informed future studies, just as the studies inform the outputs. This field, then, is a complex and shifting bricolage, with familiarity and newness, learning and adapting, and a redefinition or at least extension of how the field and leaving the field has conventionally been viewed. It has never been clear cut, and there have been ongoing connections, returns and obligations, but for me the opportunities and, importantly, attached funding of the impact agenda have enabled a 'no exit' approach so far. In the concluding section, I will reflect more critically on being in an extended field, but before this I will outline some examples of staying in the field to consider issues of positionality, familiarity and emotion in this evolving field, drawing on the framings of déjà vu and jamais vu.

Déjà vu et jamais vu

Déjà vu, already seen, has become an everyday term for encountering a situation as familiar, even though it has not been directly experienced before. Strictly speaking, these feelings of familiarity occur without the ability to identify of their source; but here I modify the term to include instances where the source of alignment can be recognised. For example, in my initial entering of the field in the first research project (Mannay et al. 2015; 2017), the accounts of children and young people that I had never met before had echoes of familiarity, reverberations of stories that had been told before.

Of course, each interaction is different, and in ethnographic fieldwork decisions to leave are often predicated in relation to the question, 'Are there phenomena that I am now taking for granted and need to re-visit?' (Delamont and Atkinson 2021: 139). This was not a feeling of over familiarity with the present, but a déjà vu based in the historical. When children talked about the painful processes of being made visible by being taken out of live classes in school to see their social worker, and of the low academic expectations that were ascribed to them because of their 'looked after' label, their experiences mapped onto generations of children who had encountered the care system. The review of the literature conducted as part of the study demonstrated the same patterns: many excellent studies had been conducted but the issues raised, and the recommendations offered, had not necessarily translated into changes in policy, practice or the everyday experiences of the care-experienced community.

Aligning with the voice of the young person discussed in the previous section, 'I didn't come for you, you come and ask questions and go, nothing ever changes', the literature review suggested that the transformations that were hoped for had not yet materialised. This déjà vu, of the already seen, contributed to a commitment to stay in, or perhaps stay with, the field. It was important that there was at least an attempt for this work not to join the shelf, where other researchers would come and conduct a new review, and then experience this same déjà vu when they worked with children and young people within the care system.

This déjà vu meant that I did not want to exit the field, but the sister term 'jamais vu' is perhaps more useful to reflect on my experiences within this extended field. Jamais vu speaks of encountering something that seems totally unfamiliar because there is too little connection between long-term memory and perceptions from the present, but here I revise this to consider not a lost memory, but one that has been deadened, managed or supressed in the field.

In the lobby of the Wales Millennium Centre, where we were gathered after completing The Fostering Network's annual fundraiser, the Foster Walk, a voice across the room shouted 'Dr Dawn'. I know that I am a doctor, the letters 'Dr' feature on a plaque on my office door, yet this calling out in this moment was unwelcome, and totally unfamiliar, because in this field I did not use this title, it was submerged, rejected and detached from my presentation of self (Goffman 1990). It was an incongruent jamais vu in the 'present' of this situational context of the event that I attend every year, which is one aspect of remaining in, or at least involved with, the field.

The caller was a young man whom I had worked with in different contexts, filling envelopes to send all schools in Wales the #messagestoschools charter designed by care-experienced young people, visiting the Waterfront Museum in Swansea as part of a youth engagement project, in the audience at different events and eating together as a group at a popular chain restaurant where they served frozen mashed potatoes. In all these encounters, I went by my first name; working across projects in this field, I am always just Dawn.

The 'positional self and prior experiences can influence the emotional self within the research journey' (Hodges 2018: 49), and the 'jamais vu' is not a problem of connections in my memory but a deliberate severing. I have worked in this field without a title because of its connotations. Of course, I have friends and colleagues who share the title, whom I like and respect. However, as a new doctor, teaching for an organisation in various venues, I was horrified by other lecturers who would announce themselves as Dr X. I watched them self-importantly demand that their room was opened and complain about any little thing to the person on the reception or security desk. Therefore, when I would go to these desks myself and politely ask how I would find the correct room, it was a relief to hear the person behind the desk say, 'ah, the Dr is not here yet for that class'. I would then have to explain that I was in fact teaching the class, but I was happy not to be recognised as that 'sort of Dr'.

In the weeks leading to The Foster Walk event, I had been working on a project with foster carers and young people (Mannay et al. 2021), where I had been again reminded of the problematic issues associated with the 'Dr X' stereotype. One foster carer had reflected on taking part in my project, which they were happy to do because I was not like the 'other one'. The 'other one', a researcher who had tried to engage them to take part in a study, had apparently begun the conversation with 'well I am Dr X and …', which was retold using a caricature of an upper-class accent. 'Well, that was it,' the foster carer told me, they definitely were not going to take part. They knew I was a Dr from the information sheet and consent form, and perhaps an internet search too, but I was told that I had been 'down to earth'.

This foster carer was also in the lobby, when '*Dr Dawn*' was shouted across the space. The extra-familiar thing of my title seemed totally unfamiliar, an unwelcome jamais vu. Madden (2019 n.p.) discusses how researchers have a 'cover story' to explain themselves in the field: the 'cover' does not mean covertness, rather, it is a way of presenting the self and the reason for entry. In coming into and remaining in the field, I have been careful to explain where I am from and my roles, which have shifted between researcher, collaborator, trainer, facilitator and messenger, in activities that are focused on data gathering, project organisation and impact. My approach to building rapport with participants has been to avoid using a title that could be read as hierarchical and to communicate that, rather than being the expert, I am hoping to learn from those whom I encounter as 'experts by experience' (Preston-Shoot 2007).

For Coffey (1999: 22), 'who is a stranger or a member, an outsider or an insider, a knower or an ignoramus is all relative and much more blurred than conventional accounts might have us believe', and my uncomfortable uncomfortableness (Hallett 2015) was an artefact of my own subjective understanding of the moment of jamais vu. When I walked over in response, the young man was laughing. I could not have a direct insight into the cause of the laughter; maybe it was amusing that the young man had bumped into me again, perhaps he called me 'Dr' because I never use the title myself in my 'cover story', possibly there was some novelty in knowing someone who is a doctor, one who is not a general practitioner and no good in a medical emergency. The rationale was no longer important, because the laughter was reassuring in itself. This is because it seemed to position me as someone who has become familiar in the field and, in having not yet left and being part of this fundraiser, not a 'researcher as a vampire' (Ward 2015: 170) attempting to 'escape with the loot, never to be seen again' (Gobo 2008b: 306) or a 'Dr' that cannot be laughed at or laugh with care-experienced young people. However, I am also a 'Dr', and this in itself is a personal difficulty for managing a balance between research, impact, responsibility and commitment, which will be the focus of the next section.

Concluding thoughts

For Pole and Hillyard (2017: 109), the 'researcher never entirely leaves the field' and 'the field never entirely leaves the researcher'. This may be true of all studies to some extent, but the bricolage of research, impact, advocacy, writing, policy work and multimodal engagement across seven years creates a different type of field. It has been a privilege to work in this field, but it has also been a challenge. This work began with different colleagues who have been a part of this journey in different ways, but for research staff it is particularly difficult to commit to the in-betweenness of funded projects when their work hours need to be accounted for in individual research centres. Additionally, although there is often funding for impact work, there may not be a relative commitment to support the time that it takes to effectively engage in impact activities.

Delivering a four-hour workshop to social workers based on the messages from care-experienced children may mean a 400-mile round trip, while mailing out a charter developed by young people to every school in Wales involved sticking lots of envelopes, even with the help of a committed group of care leavers. Watching young people enter a meeting carrying bags from the food bank, speaking to a young person on the phone who does not have heating or enough food in the COVID-19 lockdown, and listening to a kinship carer tell you that their windows were smashed while they were inside their home, is a reminder of why you have not left the field, but it is emotionally draining.

In a footnote to their chapter about working in a long-term collaborative project with care-experienced young people, as research assistants with other work commitments, Staples et al (2019: 207) comment:

> Displaying and managing patience, compassion, warmth and calmness are all part of facilitating CASCADE Voices. This emotion-work has been argued to be expected of and practised by women to a greater extent than men in academic research but not valued in the same way as other aspects of research

(Reay 2004). The gendered division and misrecognition of emotional labour is not the domain of this chapter, but it is noted because we are seeking to provide a reflexive account of the group and its functioning.

This is an important point, because this type of field engagement will not be continually supported by research time or funding; it needs to be fitted in and negotiated, and even with the impact agenda, the incremental changes that can be achieved in the lives of individuals are not necessarily the measurable outcomes favoured by academic institutions. As such, I have been fortunate to be able to stay with this field, but it has had associated costs. In my early research, as a later entrant to university, one of my first projects on inequalities was lost, and did not come to fruition as a doctoral study (Mannay 2019). However, the insights gained here in relation to uneven playing fields have been an ongoing feature in all my work to date. Accordingly, although I am not planning to 'leave the field' if there are changes ahead, this research and the liaisons built in the field will have an enduring impact on how I view research, relationality and responsibility.

Note

1 The system for assessing the quality of research in UK higher education institutions.

References

Atkinson, P. and Coffey, A. (2002). 'Revisiting the relationship between participant observation and interviewing', in J.F. Gubrium and J.A. Holstein (eds) *Handbook of Interview Research*. pp. 801–814. Thousand Oaks, CA: SAGE.

Barnes, V., Clouder, D., Pritchard, J., Hughes, C. and Purkis, J. (2003). 'Deconstructing dissemination: Dissemination as qualitative research'. *Qualitative Research*. 3(2): 147–64.

Bloor, M. and Wood, F. (2006). 'Leaving the field', in M. Bloor and F. Wood (eds) *Keywords in Qualitative Methods*. pp. 111–112. London: SAGE.

Bourdieu, P. (1990). *The Logic of Practice*. Stanford, CA: Stanford University Press.

Coffey, A. (1999). *The Ethnographic Self: Fieldwork and the Representation of Identity*. London: SAGE.

Coffey, A. (2018). *Doing Ethnography: The Sage Qualitative Research Kit*. (2nd edn). London: SAGE.

Delamont, S. (2012). 'Milkshakes and convertibles: An autobiographical reflection'. *Studies in Symbolic Interaction*. 39: 51–69.

Delamont, S. and Atkinson, P. (1995). *Fighting Familiarity: Essays on Education and Ethnography*. Cresskill, NJ: Hampton Press.

Delamont, S. and Atkinson, P. (2021). *Ethnographic Engagements: Encounters with the Familiar and the Strange*. Abingdon: Routledge.

Fine, G.A. and Deegan, J. (1996). 'Three principles of Serendip: Insight, chance, and discovery in qualitative research'. *Qualitative Studies in Education*. 9(4): 434–447.

Gobo, G. (2008a). *Doing Ethnography*. London: SAGE.

Gobo, G. (2008b). 'Leaving the field', in G. Gobo. (ed.) *Doing Ethnography*. pp. 306–313. London: SAGE.

Goffman, E. (1990 [1959]). *The Presentation of Self in Everyday Life*. London: Penguin.

Hallett, S. (2015). 'An uncomfortable comfortableness: "Care", child protection, and child sexual exploitation'. *British Journal of Social Work*. 46(7): 2137–2152.

Hammersley, M. and Atkinson, P. (2007). *Ethnography: Principles in Practice*. (3rd edn). London: Taylor and Francis.

Hodges, A. (2018). 'The positional self and researcher emotion: Destabilising sibling equilibrium in the context of cystic fibrosis', in T. Loughran and D. Mannay (eds) *Emotion and the Researcher: Sites, Subjectivities, and Relationships. (Studies in Qualitative Methodology, Volume 16)*. pp. 49–63. Bingley: Emerald.

Hopkins, P. and Noble, G. (2009). 'Masculinities in place: Situated identities, relations and intersectionality'. *Social and Cultural Geography*. 10(8): 811–819.

Hugman, R., Pittaway, E. and Bartolomei, L. (2011). 'When "do no harm" is not enough: The ethics of research with refugees and other vulnerable groups'. *British Journal of Social Work*. 41(7): 1271–1287.

Kara, H., Lemon, N., Mannay, D. and McPherson, M. (2021). *Creative Research Methods in Education: Principles and Practices*. Bristol: Policy Press.

Knowles, C. and Burrows, R. (2014). 'The impact of impact'. *Etnográfica*, 18(2): 237–254.

Lincoln, S. (2012). *Youth Culture and Private Space*. Basingstoke: Palgrave Macmillan.

Lisiak, A. and Krzyżowski, Ł. (2018). 'With a little help from my colleagues: Notes on emotional support in a qualitative longitudinal research project', in T. Loughran and D. Mannay (eds) *Emotion and the Researcher: Sites, Subjectivities, and Relationships. (Studies in Qualitative Methodology, Volume 16)*. pp. 197–212. Bingley: Emerald.

Loughran, T. and Mannay, D. (2018). 'Introduction: Why emotion matters', in T. Loughran and D. Mannay (eds) *Emotion and the Researcher: Sites, Subjectivities, and Relationships. (Studies in Qualitative Methodology, Volume 16)*. pp. 1–18. Bingley: Emerald.

Madden, R. (2019). 'Entering the field', in P. Atkinson, S. Delamont, A. Cernat, J.W. Sakshaug and R.A. Williams (eds) *SAGE Research Methods Foundations*. London: SAGE. www.doi.org/10.4135/9781526421036769642.

Mannay, D. (2019). 'What happens when you take your eye off the ball? Reflecting on a "lost study" of boys' football, uneven playing fields and the longitudinal promise of "esprit de corps"', in R.J. Smith and S. Delamont (eds) *The Lost Ethnographies: Methodological Insights from Projects that Never Were. (Studies in Qualitative Methodology, Volume 17)*. pp. 103–112. Bingley: Emerald.

Mannay, D. and Morgan, M. (2015). 'Doing ethnography or applying a qualitative technique? Reflections from the "waiting field"'. *Qualitative Research*. 15(2): 166–182.

Mannay, D., Rees, A. and Roberts, L. (eds) (2019). *Children and young people 'looked after'? Education, intervention and the everyday culture of care in Wales*. Cardiff: University of Wales Press.

Mannay, D., Staples, E., Hallett, S., Roberts, L., Rees, A., Evans, R. and Andrews, D. (2015). *Understanding the Educational Experiences and Opinions, Attainment, Achievement and Aspirations of Looked After Children in Wales*. Project Report. Cardiff: Welsh Government.

Mannay, D., Evans, R., Staples, E., Hallett, S., Roberts, L., Rees, A. and Andrews, D. (2017). 'The consequences of being labelled "looked-after": Exploring the educational experiences of looked-after children and young people in Wales'. *British Educational Research Journal*. 43(4): 683–699.

Mannay, D., Staples, E., Hallett, S., Roberts, L., Rees, A., Evans, R. and Andrews, D. (2019). 'Enabling talk and reframing messages: Working creatively with care experienced children and young people to recount and re-represent their everyday experiences'. *Child Care in Practice*. 25(1): 51–63.

Mannay, D., Smith, P., Turney, C. and Davies, P. (2021). '"Becoming more confident in being themselves": The value of cultural and creative engagement for young people in foster care'. *Qualitative Social Work*. 21(3). https://doi.org/10.1177/14733250211009965 (accessed 12 December 2022).

Pole, C. and Hillyard, S. (2017). 'When it's time to go', in C. Pole and S. Hillyard (eds) *Doing Fieldwork*. pp. 107–122. London: SAGE.

Preston-Shoot, M. (2007). 'Whose lives and whose learning? Whose narratives and whose writing? Taking the next research and literature steps with experts by experience'. *Evidence & Policy: A Journal of Research, Debate and Practice*. 3(3): 343–359.

Reay, D. (2004). 'Cultural capitalists and academic habitus: Classed and gendered labour in UK higher education'. *Women's Studies International Forum*. 27(1): 31–39.

REF (2011). 'Assessment Framework and Guidance on Submissions'. REF 02.2011. www.ref.ac.uk/pubs/2011-02/.

Sheppard, L. (2018). 'Poor old mixed-up Wales: Entering the debate about bilingualism, multiculturalism and racism in Welsh literature and culture', in T. Loughran and D. Mannay (eds) *Emotion and the Researcher: Sites, Subjectivities, and Relationships. (Studies in Qualitative Methodology, Volume 16)*. pp. 197–212. Bingley: Emerald.

Smith, R.J. and Delamont, S. (eds) (2019). *The Lost Ethnographies: Methodological Insights from Projects that Never Were. (Studies in Qualitative Methodology, Volume 17)*. Bingley: Emerald.

Staples, E., Roberts, L., Lyttleton-Smith, J., Hallett, S. and CASCADE Voices. (2019). 'Enabling care experienced young people's participation in research: CASCADE voices', in D. Mannay, A. Rees and L. Roberts (eds) *Children and Young People 'Looked After'? Education, Intervention and the Everyday Culture of Care in Wales*. pp. 196–209. Cardiff: University of Wales Press.

Timmins, F. (2015). 'Disseminating nursing research'. *Nursing Standard.* 29(48): 34–39.
Ward, M.R.M. (2015). *From Labouring to Learning: Working-class Masculinities, Education and De-industrialization.* Basingstoke: Palgrave Macmillan.
Welsh Government. (2016). *Raising the Ambitions and Educational Attainment of Children Who Are Looked After in Wales.* Cardiff: Welsh Government.
Welsh Government. (2017). *Making a Difference: A guide for the Designated Person for Looked After Children.* Cardiff: Welsh Government.
Wilson, A. (2021). 'Reading and Resisting Racialised Misrecognition: An Exploration of Race Scripts, Epiphany Moments and Racialised Selfhood'. PhD thesis, Cardiff University.

8

Student voices 'echo' from the ethnographic field

Janean Robinson, Barry Down and John Smyth

Introduction

> If the heuristic revisit moves forward in time, from the earlier study to the later one that it frames, the archaeological revisit moves backward in time to excavate the historical terrain that gives rise or gives meaning to the ethnographic present. (Burawoy 2003: 671)

When we were invited to contribute to this collection on 'ethnographic exits' in 2019 we stumbled across e-mail archives from our Australian Research Council (ARC) research project (Down and Smyth 2010) tracing thirty-two young people over a three-year period as they transitioned from school to work. These archives triggered memories of the power and timelessness of student narratives and how they continue to 'speak' from the past; in particular, responses received from Jacinta and Lucas,[1] two participants who profoundly influenced our thinking.

As critical ethnographers, we never really *leave* the field. We have a commitment and responsibility to honour the lives of our participants. We cannot help but wonder how these young people are faring and ask questions like, 'Where are they now? Are they okay? Do they have work and families?' The very fact that we are still engaged suggests something more profound is going on here. There are genuine dilemmas in 'letting go' at both personal and intellectual levels, and yet we remain deeply attached in ways that we do not yet fully comprehend and feel some connection and desire to 'check out' how things are 'going on' for them. For this reason, we sent each of our participants a short message in 2014, a year after the completion of our research:

> Hello ... you may recall that this time last year we had our third and final interview for the ARC Linkage Project: 'Getting a job'. Even though the formal interviews are complete we would be very keen to find out how things are going for you. What are you doing now? Please do not hesitate to call on the

number (below), or if you prefer please send a text message or email. We look forward to hearing from you soon. (e-msg sent to participants of ARC Project, March 2014)

We believe these kinds of lingering 'entanglements' deserve greater scrutiny in terms of understanding the methodological implications of researching young lives (Matthiesen 2020: 2); especially those most marginalised because of multiple and complex circumstances beyond their control, including shifts in the global economy (Harvey 2007). As Giroux (2009) explains it, 'youth are now viewed as either consumers, on the one hand, or as troubling, reckless, and dangerous persons, on the other' (Giroux 2009: 3). Consequently, the intensity of our commitment and advocacy grows rather than diminishes with time, because the further we appear to move 'away' from the actual fieldwork, the closer we become. We acknowledge and value the generosity of the young people who so willingly shared parts of their lives with us, no matter how partial or incomplete it may seem. We are aware of the privilege we have been afforded, and with this comes a sense of responsibility which we could not easily dismiss.

These kinds of human sensibilities are not always welcome in the context of the neoliberal university where researchers are urged to move rapidly from one research grant to the next in a punishing environment of 'publish or perish'. The obsession with metrics (e.g., research income, citations and global rankings) as the means and ends of scholarly worth serves only to diminish the value of critical social research in the academy (Giroux 2014; Smyth 2017). Despite these pressures, we find ourselves stuck in the existential realities of these stories and how we might respond ethically, morally and pedagogically.

Neoliberalism, exclusion and the traces of young lives

Inclusive education is everybody's business. (Slee 2011: 83)

In his book, *The Terror of Neoliberalism* (2004), Henry Giroux argues that neoliberalism is not just about economics but also something much more sinister culturally and politically. In his view, neoliberalism connects us to global fields of existence (our values, our ideologies and our political policy practices), all the while reconfiguring and displacing democratic values and public institutions in the interests of global capitalism. As McMurtry explains so well, neoliberalism is the 'cancer stage of capitalism' (McMurtry 2012). It creates a dire threat to human existence environmentally, politically and socially, characterised by the 'dispossessing effects' of market prescriptions such as growing levels of unemployment, piecemeal and poorly paid casual jobs, loss of social benefits and loss of economic power (McMurtry 2012: 54). Nowhere is this 'theatre of cruelty' (Giroux 2009) or 'wasted lives' (Bauman 2004) more apparent than in the lives of young people who become 'collateral damage' (Bauman 2011) or disposable commodities in a ruthless Darwinian game of 'survival of the fittest'. In the context of precarity and stark levels of inequality (Wilkinson and Pickett 2009), matters of identity and citizenship (Hall, Coffey and Williamson 1999), founded on principles of full freedom and basic income security for young people, become pivotal issues in democratic societies (Standing 2009).

As we delved into broader structural issues, we comprehended how they impacted on the lives of young people themselves. Under neoliberalism, schools are increasingly aligned to the imperatives of the economy, leading to narrowly conceived and instrumentalist approaches to teaching and learning. Sahlberg (2010) coined the term Global Education Reform Movement (GERM) to describe a suite of policy manoeuvres borrowed from the business world to reform education, including competition, accountability, managerialism, privatisation, choice and standardisation. At the school level, this emphasis is translated into a limiting set of core subjects, prescribed curriculum, didactic teaching and high-stakes testing (Lipman 2004). Examples include PISA (Programme for International Student Assessment) and NAPLAN (National Assessment Program for Literacy and Numeracy). It is hardly surprising therefore to find students 'dropping out, drifting off, and being excluded' (Smyth and Hattam 2004) from school in alarming numbers. What we are witnessing in the neoliberal policy landscape is escalating levels of educational and social inequality (Connell 1993; Youdell 2004). To help us better understand the fallout from this broader social and economic context we ethnographically tracked the respective journeys of thirty-two young people as they transitioned from school to work. We wanted to listen to what they themselves had to say about school and the barriers they faced in finding meaningful work. Drawing on their experiences, we identified the kinds of educational policies and practices that assisted them in achieving their imagined futures (Down, Smyth and Robinson 2019). This cultural and pedagogical work is grounded in a deep commitment to the moral, ethical and social purposes of education (Dewey 1963) and what it means to be truly educated (Smyth and McInerney 2014). In the next section, we share the narratives of two young people, Jacinta and Lucas, who represent how neoliberalism excludes far too many from productive and meaningful lives.

Narratives that echo

As storytellers we focus on the narratives of young people to help us understand the complexities of social life. This kind of research reveals a great deal about the pain and suffering of marginalised young people, often silenced and made 'invisible' within a larger society (Giroux 2009: 4). For critical researchers this has haunting effects, as it presents significant emotional, intellectual and pedagogical challenges that cannot be readily swept under the carpet. As Smyth and McInerney (2013) argue, these stories must be respected by showing a willingness 'to confront problematics' and 'taking a genuine democratic stand for the subjects of their research who are treated unjustly' (Smyth and McInerney 2013: 2). The authors explain:

> It thus has a fundamental and unswerving commitment to re-assembling, reconstructing, and portraying accounts of social life in ways that honour its inherent complexity– rather than purporting to be able to render it down to fragments, 'bottom lines', 'recommendations' or meaningless metrics (Smyth and McInerney 2013: 3).

In our own collaborative research, we find that personal relationships and experiences (Clandinin and Connelly 1988) are central to getting up close to the lives of participants in ways that make it difficult to 'let go'. We continue to wonder where they are now and how they are doing. Smyth shares this sentiment in an e-mail reflection:

> For me, 'leaving the field' is not so much to do with the issues of trying to keep in contact with the young informants as personages – for that would for most purposes be a practical impossibility – but rather, it is a case of how their stories continue to haunt me, in the sense of demanding that I return to them to see if I can make further sense of them, with the passage of time to reflect. I often wonder if I have done them the full justice they deserve in this regard, and I always feel conflicted about it. (Personal e-mail correspondence, 6 November 2019)

This feeling of being haunted by past encounters returns in unexpected places and moments as we endeavour to create socially just forms of research capable of constructing a more democratic vision and practice of education (Riddle and Apple 2019). Even though we move on and become immersed in other projects, there are always those narratives that continue to 'reappear' within our writing (Robinson, Down and Smyth 2018) in ways that challenge, interrupt and 'shake up' the dehumanising effects of neoliberal practices on young people's lives. While these narratives are contextual, they also speak back to a broader set of policy trajectories, and, in this sense, the stories echo 'for' and 'on behalf of' marginalised young people in different contexts.

Arnot and Reay (2007) refer to '*a sociology of pedagogic voice*' because it engages methodologically and theoretically with power relations to explain individual experience (Arnot and Reay 2007: 312). By way of example, we briefly allude to two stories that continue to 'haunt' us in different ways. First, Lucas, a young boy who believed in the promise of schooling as a way out of intergenerational unemployment in his family and community. He was told that if he studied hard and completed a series of vocational programmes and certificates he would be rewarded with an electrical apprenticeship. Sadly, the harsh realities of the labour market dashed his dreams. His sense of disillusionment and frustration were palpable. Drawing on Lucas's experience, and invoking Mills (1971), we seek to understand how 'private matters' are indeed 'public issues'.

Second, Jacinta's story reveals a great deal about the complexity and messiness of young lives and the interferences they face in 'hanging in' at school. Jacinta found school to be an alienating experience characterised by conflict and resistance which ultimately led to her exclusion. For Jacinta, school was boring and largely irrelevant, because teachers showed little empathy towards her needs, interests or desires (Robinson, Down and Smyth 2018; 2019).

The stories shared by both Lucas and Jacinta assisted in our struggle to *reimagine* what a more democratic education might look like. What strikes us most about Lucas and Jacinta is their sense of determination, resilience and courage in the face of multiple obstacles and barriers in their young lives. Lucas, for example, was living in a region where there was 20 per cent youth unemployment. In his words:

> Since leaving school I haven't really done much. I've just been sitting at home doing nothing. Now I spend an hour in the morning looking for jobs or apprenticeships. I just have a quick look to see if anything new has popped up and if there is anything I will apply for it.
>
> In the electrical you might get one pop up every week for the whole of Western Australia. Apprenticeships are just going down like that and I've no idea why. Employers are looking for experienced workers. So, you need experience to get a job, but you need a job to get experience!
>
> You get a lot of things disappearing because they may see someone with a better-quality resume or cover letter. The problem is you're trying to get ahead of everyone else but everyone else is also trying to get ahead. Usually they'll just get like the top ten best ones and then bring them in for an interview. It's tough getting an interview. When I go for the interview normally I'm like the third or the second best. Someone always just beats me. When they say, 'Just not the best', I'm wondering what else they want. (Transcript from third interview, March 2013)

Like Lucas, Jacinta also challenged us to think more deeply about the complexity and messiness of young lives. Although she had 'signed off' in a text message in 2014, she still wanted to 'keep in contact'. Unfortunately, this did not occur, for reasons we may not fully understand, even though we are acutely aware of the complex social, family and school circumstances in which she found herself.

Jacinta's story is complicated, but it serves to remind us of the kinds of barriers young people face as they seek to navigate their way in the adult world, as expressed in an extract taken from our second interview with Jacinta:

> The school psychiatrist and the nurse have also helped me quite a bit. I can talk to them. A couple of teachers helped me. There are only two nice teachers in the whole school. I like them because they don't judge you. I put up with the others. I never used to be able to deal with teachers I didn't get on with. I used to get really angry and swear and throw things around the room, but I have quietened down a bit. My friends have helped me to work through some issues and I had to do an anger management course at 'The Link'. Some things at home have made it difficult for me. When I first started school here my mum was going to jail. I was hanging out with the wrong people. I hardly came to school because I didn't want to be here. Life at home wasn't good and there was the usual fighting. (Transcript from second interview, May 2012)

In this short recollection, Jacinta expresses her own self-awareness in needing to act sooner rather than later so as not to 'end up in the bottom of the heap'. Schools like the one Jacinta attended find it difficult to accommodate students who are 'living on the edge' (Smyth and Wrigley 2013) while dealing with a range of complex issues including poverty, violence, unemployment, drug and alcohol abuse and broken relationships. They simply expect students to go back to class, comply, be quiet and absorb the prescribed curriculum, irrespective of their daily struggles. Someone less determined and self-motivated than Jacinta could easily fall over again. This left us facing the dilemma of how best to advocate on behalf of young people like Jacinta. Fitzpatrick (2019: 172) shares similar concerns about her own research participants, who were transitioning from school to post-school life:

> I wonder in hindsight if they felt like they had let me down and I wonder now about the ethics of researching their lives at a time they could enact little control over their financial situations and life choices ... I am wondering now, as I write this chapter, what might be possible if I return to this project 10 years later?

Jacinta did not have the cultural capital (Bourdieu 1971) to navigate the school system at a time when things were not going well for her in the context of an unstable family life; especially after her chequered school experiences of being suspended and shunted from school to school when home was so unpredictable. Jacinta's text message in Figure 1 reminds us of how life circumstances can be very different for different classes of students (Reay 2017).

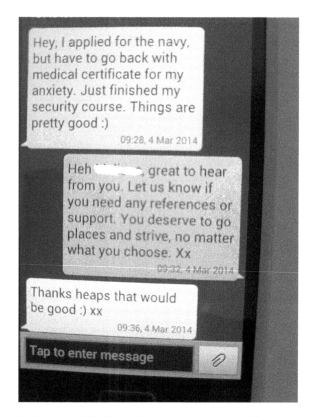

Figure 1 Segment of SMS text between Robinson and Jacinta (2014)

As Smyth and Simmons (2018) explain, 'class is not fixed or defined by category but rather something experienced through relationship' (Smyth and Simmons 2018: 2). Nowhere is this more apparent than in Jacinta's story. Given her complicated mental health issues and family circumstances, she found the demands of schooling difficult to navigate. There seemed to be little understanding or support from teachers willing to assist her through these difficult days. Jacinta identified the impact of these struggles in a text message we received from her stating 'Hey, I applied for The Navy, but I have to go back with a medical certificate for my anxiety'. Jacinta had been informed that she had to gain medical clearance before she could apply for entrance into the Australian navy. Labelling young people in deficit ways – 'troublesome', 'remedial' or 'at risk' – can have lasting effects on their imagined futures. As Greene (2001: xvi) points out, these labels 'carry the messages of power: they demean, they exclude; they create stereotypes'. Despite these obstacles, Jacinta showed remarkable resilience and courage. She completed a tourism course at school as she endeavoured to rewrite her identity and 'escape' the limitations of her home and school life. Hall et al. (1999: 503) explain that these identity transitions 'leave young people ambiguously placed in respect of an adult status'. According to them, 'some [are] able to take advantage of this extension of youth', while others, like Jacinta and Lucas, 'may find it a frustrating time' because the sense of themselves and their place in society 'seems indeterminate and unresolved' (Hall et al. 1999: 503). In this context, Slee (2011) argues that as educators we should be asking serious questions 'about the power relations of schooling' (Slee 2011: 155) and the implications of excluding students from the benefits of education. Bauman (2004: 132) explains how these exclusionary practices work:

> Spotting the people who 'do not fit' into the place they are in, banishing them from that place and deporting them 'where they belong', or better still never allowing them to come anywhere near in the first place.

When one locates the narratives of young people like Lucas and Jacinta in the context of broader social and political arguments, we then assume a greater ethical and political responsibility to change the way things are. Knowing their stories makes it harder to ignore them. This is the emancipatory potential of critical ethnography. It calls on researchers to listen deeply to what young people have to say and to represent their stories as opportunities for social transformation. In the next section we elaborate on what this means for our own research and reflexivity.

Theory and critical reflexive ethnography

We take as our starting point the view that evidence garnered from the field can only be properly comprehended in dialogue with theory. Ball (2006) argues that theory can be too easily dismissed when it is 'either spoken in hushed tones before speakers move on to present their 'findings', or it is some kind of add-on or aside that does little' (Ball 2006: 1). In addressing this shortcoming, we once again draw on Mills' (1971) 'sociological imagination', whereby individual experiences are located in the context of the social world and the broader structural conditions which maintain the way things are. We believe that theory must be taken seriously when exploring participants' lives. Smyth and McInerney (2013: 3) elaborate:

> we all carry theories of one kind or another and those theories are worked on and shaped as a result of our field experiences, and in turn, our encounters with the field are reshaped by the theories we bring to our research ... and indeed ... it is through this organic and iterative relationship that knowledge gains its vitality and vibrancy.

Lather's (1986) notion of 'dialectic theory building', whereby 'data constructed in context are used to clarify and reconstruct theory' (Lather 1986: 267), is especially helpful. Lather argues that research is a process of shuffling between ideas and everyday experiences. In a similar way, Willis and Trondman (2000) identify theory as one of the crucial features of ethnography, as it is a 'precursor, medium and outcome of ethnographic study and writing (Willis and Trondman 2000: 7). But, equally important, it must be critical in terms of 'recording and understanding lived social relations … from the point of view of how they embody, mediate and enact the operations and results of unequal power' (Willis and Trondman 2000: 10).

In this context, Burawoy (2003) advocates the importance of 'revisiting' evidence from the field as a means of transporting it from its unconscious past into a historicised world' (Burawoy 2003: 646). We like the way Burawoy draws attention to the relevance of data 'over time' (Burawoy 2003: 646). Atkinson and Morriss (2017) argue that, as critical ethnographers, we should be preoccupied not 'with knowledge acquisition per se' but, rather, with 'processes by which such learning is achieved'. This approach to research enables us to 'maintain reflexive self-awareness and analytic distance' (Atkinson and Morriss 2017: 329). Robinson (2014: 207) sheds insight on the process of reflexive knowledge production and self-awareness while 'reshaping' data over time and analytical distance:

> The voices of my subjects linger in my head, and yet there is a void. The interviews are finished, so I can no longer share or communicate with them. I am left instead with a recording and a collection of their words on my paper that I tinker and toil over (Journal entry, September 2007).
>
> It is difficult to know how much to allow yourself to represent *the other* and how much to allow yourself to *feel*, knowing that you are going to have to *(un)feel* again and continue with further research and other work which may or may not involve 'those' individual participants. (Journal entry, February 2008)
>
> Now I am using these 'scripts' 10 months later and feel a sense of something – I don't know what words to use, but it is an acknowledgement, an awareness, that I probably will never see these persons again, and yet they have provided so much insight; these 3 interviews into 'their' worlds. In creating this dialogue and reliving the actual event, is like being able to step 'back' inside again, being privy to my participants' sacred spaces; their combined interactive knowledges and shared experiences. (Journal entry, April 2008)

While we cannot always solve these research 'dilemmas', it does allow us to 'turn back' (Davies 2008: 4) to re-examine one's fieldwork and the kind of meaning-making it generates. In this sense, we are no longer haunted or 'stuck' wondering about our participants. Instead, we have a way of leaving the field with a view to theorising 'knowledge-in-action' (Atkinson and Morriss 2017) and persistently reshaping our theory and methodology. This allows us to create spaces and places for participants 'to speak back' to dominant discourses by recirculating their experiences as part of ongoing research endeavours in the context of contemporary events and circumstances.

Conclusion: does one ever really leave the field?

> Projects are never finished or entirely lost, even whilst aspects disappear over the edge. (Fitzpatrick 2019: 173)

This chapter has been a composition of memory; a peregrination in sharing why some stories echo from the field and why it matters. We believe the stories of young people like Lucas and Jacinta are just as relevant today as when we first heard them. They continue to inform our lives, our teaching and our research. As Smyth and McInerney (2013: 17) explain:

> as researchers we have a moral and ethical responsibility beyond the 'thin' imposed views of university ethics committees – to work *with* and advance the lives of those who are institutionally and systematically the most excluded and silenced.

As critical ethnographers we have a moral, ethical and political responsibility to work *with* and advocate *for* the lives of the least advantaged (Connell 1993). In pursuing this task, we conclude by offering some reflexive insights into our experience of 'exiting the field'. For us, the process of exiting the ethnographic field:

- acknowledges that 'one never really *leaves* the field'!
- maintains a critical self-reflexivity over time and place;
- generates meaning from interpersonal relationships and 'purposeful conversations' (Burgess 1988);
- portrays the lives of marginalised and silenced voices;
- reframes taken-for-granted explanations of the way things are;
- interrupts orthodox beliefs, rituals, assumptions and practices of everyday life;
- theorises issues by moving beyond deficit and victim-blaming discourses;
- recognises the moral, ethical and political nature of knowledge production *for* and *with* informants;
- views researchers/participants as 'connoisseurs' of 'knowledge-in-action' (Atkinson and Morriss 2017: 329–330);
- understands the constant 'reformulation of ethnographic imaginings' and 'a curiosity within an overall theoretical sensibility' (Willis 2000: 114).

We believe these kinds of theoretical and methodological orientations allow us to maintain an ongoing dialogue with our participants' lives. Drawing on this rich archive of knowledge we are better placed to undertake the challenge of making 'the physical, social, cultural and educational arrangements of schooling better for all' (Slee 2011: 13). This is no easy feat; but, as Giroux (2009) warns us, there is a need to act quickly, both politically and pedagogically, to educate 'critically engaged social agents capable of addressing the meaning, character, fate, and crisis of democracy' (Giroux 2009: 181).

In this chapter we have shared our experience of utilising a spirit of criticality in our ethnographic work with marginalised young people and why it matters. As critical educators we are deeply committed to the principles of democracy and social justice and have an emotional attachment to our participants as part of a broader social movement. At the same time, we also know that eventually we will have to '*let them go*'. Even so, the narratives of young people like Lucas and Jacinta will continue to echo from the past as we expand our thinking, research and writing based on their generosity and wisdom. Ultimately, the purpose of our work is to generate new forms of consciousness and activism among educators, young people and the wider community, with a view to creating a more socially just world.

Note

1 We have used pseudonyms for our participants throughout this chapter

References

Arnot, M. and Reay, D. (2007). 'A sociology of pedagogic voice: Power, inequality and pupil consultation'. *Discourse: Studies in the Cultural Politics of Education*. 28(3), 311–325.

Atkinson, P. and Morriss, L. (2017). 'On ethnographic knowledge'. *Qualitative Inquiry*. 23(5): 323–331.

Ball, S.J. (2006). 'Introduction'. *Discourse: Studies in the Cultural Politics of Education*. 27(1): 1–2. doi: 10.1080/01596300500510203.

Bauman, Z. (2004). *Wasted Lives: Modernity and Its Outcasts*. Oxford: Polity.

Bauman, Z. (2011). *Collateral Damage: Social Inequalities in a Global Age*. Malden, MA: Polity Press.

Bourdieu, P. (1971). 'Cultural reproduction and social reproduction', in J. Karabel and A. Halsey (eds) *Power and Ideology in Education*. pp. 487–511. New York: Oxford University Press.

Burawoy, M. (2003). 'Revisits: An outline of a theory of reflexive ethnography'. *American Sociological Review*. 68(5): 645–679.

Burgess, R. (1988). 'Conversations with a purpose: The ethnographic interview in educational research'. *Studies in Qualitative Methodology.* 1(1): 137–155.
Clandinin, D.J. and Connelly, F.M. (1988). *Teachers as Curriculum Planners: Narratives of Experience.* New York: Teachers College Press.
Connell, R. (1993). *Schools and Social Justice.* Toronto. Our Schools/Our Selves Education Foundation.
Davies, C.A. (2008). *Reflexive Ethnography: A Guide to Researching Selves and Others.* London: Routledge.
Dewey, J. (1963). *Experience and Education.* New York: Collier Books.
Down, B. and Smyth, J. (2010). '"Getting a job": Identity formation and schooling in communities at disadvantage'. Australian Research Council Linkage Scheme Grant (Project ID LP100100031).
Down, B., Smyth, J. and Robinson, J. (2019). 'Problematising vocational education and training in schools: Using student narratives to interrupt neoliberal ideology'. *Critical Studies in Education.* 60(4): 443–461.
Fitzpatrick, K. (2019). 'The edges and the end: On stopping an ethnographic project, on losing the way', in S. Delamont and R.J. Smith (eds) *The Lost Ethnographies: Methodological Insights from Projects That Never Were. Studies in Qualitative Methodology.* 17: 165–175.
Giroux, H. (2004). *The Terror of Neoliberalism: Authoritarianism and the Eclipse of Democracy.* Boulder, CO: Paradigm Publishers.
Giroux, H.A. (2009). *Youth in a Suspect Society: Democracy or Disposability?* New York, Palgrave Macmillan.
Giroux, H. (2014). *Neoliberalism's War on Higher Education.* Chicago, IL: Haymarket Books.
Greene, M. (2001). 'Foreword', in G. Hudak and P. Kihn (eds) *Labeling: Pedagogy and Politics.* pp. xvi–xvii. London: Routledge-Falmer.
Hall, T., Coffey, A. and Williamson, H. (1999). 'Self, space and place: Youth identities and citizenship'. *British Journal of Sociology of Education.* 20(4): 501–513.
Harvey, D. (2007). *A Brief History of Neoliberalism.* New York: Oxford University Press.
Lather, P. (1986). 'Research as praxis'. *Harvard Educational Review.* 56(3): 257–277.
Lipman, P. (2004). *High Stakes Education: Inequality, Globalization, and Urban School Reform.* New York: Routledge.
Matthiesen, N. (2020). 'A question of access: Metaphors of the field'. *Ethnography and Education.* 15(1): 1–16.
McMurtry, J. (2012). *The Cancer Stage of Capitalism: From Crisis to Cure.* London: Pluto.
Mills, C.W. (1971 [1959]). *The Sociological Imagination.* New York: Penguin Books.
Reay, D. (2017). *Miseducation: Inequality, Education and the Working Classes.* Bristol: Policy Press.
Riddle, S. and Apple, M.W. (2019). 'Education and democracy in dangerous times', in S. Riddle and M.W. Apple (eds) *Re-imagining Education for Democracy.* pp. 1–9. London and New York: Routledge.
Robinson, J. (2014). 'Dilemmas and deliberations in reflexive ethnographic research'. *Ethnography and Education.* 9(2): 196–209.

Robinson, J., Down, B. and Smyth, J. (2018). '"Shaking up" neoliberal policy in schools: Looking for democratic alternatives in Jacinta's satchel'. *Global Studies of Childhood*. 8(4): 392–403.

Robinson, J., Down, B. and Smyth, J. (2019). 'Jacinta's story: Challenging neoliberal practices and creating democratic spaces in public high schools', in S. Riddle and M.W. Apple (eds) *Reimagining Education for Democracy*. pp. 156–273. London and New York: Routledge.

Sahlberg, P. (2010). 'Rethinking accountability in a knowledge society'. *Journal of Educational Change*. 11: 45–61.'

Slee, R. (2011). *The Irregular School: Exclusion, Schooling and Inclusive Education*. New York: Routledge.

Smyth, J. (2017). *The Toxic University: Zombie Leadership, Academic Rock Stars, and the Neoliberal University*. London: Palgrave Macmillan.

Smyth, J. and Hattam, R., with Cannon, J., Edwards, J., Wilson, N. and Wurst, S. (2004). *'Dropping Out', Drifting Off, Being Excluded: Becoming Somebody Without School*. New York: Peter Lang.

Smyth, J. and McInerney, P. (2013). 'Whose side are you on? Advocacy ethnography: some methodological aspects of narrative portraits of disadvantaged young people, in socially critical research'. *International Journal of Qualitative Studies in Education*. 26(1): 1–20.

Smyth, J. and McInerney, P. (2014). *Becoming Educated: Young People's Narratives of Disadvantage, Class Place and Identity*. New York: Peter Lang.

Smyth, J. and Simmons, R. (2018). 'Where is class in the analysis of working-class education?' in R. Simmons and J. Smyth (eds) *Education and Working-Class Youth: Reshaping the Educational Furnace*. pp. 1–28. Switzerland: Palgrave Macmillan.

Smyth, J. and Wrigley, T. (2013). *Living on the Edge: Rethinking Poverty, Class and Schooling*. New York: Peter Lang.

Standing, G. (2009). *Work after Globalization*. Cheltenham: Edward Elgar Publishing.

Wilkinson, R.G and Pickett, K. (2009). *The Spirit Level: Why Greater Equality Makes Societies Stronger*. New York: Bloomsbury Press.

Willis, P. (2000). *The Ethnographic Imagination*. Cambridge, UK: Polity Press.

Willis, P. and Trondman, M. (2000). 'Manifesto for ethnography'. *Ethnography*. 1(1): 5–16.

Youdell, D. (2004). 'Engineering school markets, constituting schools and subjectivating students: The bureaucratic, institutional and classroom dimensions of educational triage'. *Journal of Education Policy*. 19(4): 407–431.

9

Public space and visible poverty: research fields without exit

Andrew P. Carlin

Introduction

Pavements are research sites *sui generis*. This chapter discusses identifiable pavement cultures as members' methods in public space. Pavement cultures are constituted by cohorts of pedestrians within urban spaces. A key reference point is a landmark video-based study (Lee and Watson 1993), which set new parameters for doing ethnography by studying members' methods in urban social spaces. This was a trailblazing report that, by applying the conceptual insights of conversation analysis to non-talk environments, and by using pedestrian locomotion as an ordinary, perspicuous activity, sought to realign the sequential and categorial forms of conversation analysis, which were becoming increasingly distal from each other. This realignment is known as *ethnomethodology*.

Insights from Lee and Watson (1993) afforded the ethnographic analysis of practices of visible poverty, e.g., selling copies of *The Big Issue* magazine in Manchester, UK, which problematised both the reification of video-recording as a method; and begging, from a collaborative fieldwork project in a town in northern Europe. This project was an ethnographic study of begging in a public space, known locally as 'Cathedral Square' (pseudonym). These inquiries illustrate the contingencies of studying everyday life, namely, (i) you never really leave the field, and (ii) the field is always with you:

> Society is produced and reproduced all around us all the time. As persons sociologists constantly take part in the social scenes they study and they do have access to the original scenes as they unfold. Sociology as a discipline is interested in all forms of social life. Therefore, the data are everywhere. (Rawls 2009: 94)

Some of the fieldwork on visible poverty presented here was a team ethnography, and I rely heavily upon fieldnotes worked up by the team (Carlin, Evergeti and Murtagh 1999). Preserving the phenomenon of inquiry as experienced by members of society presents problems for recording technologies, which are explored within this study. Although observations are tied explicitly to particular settings, what was observed on the streets – people walking along pavements – occurs worldwide. Once noticed, the phenomenon of pedestrian locomotion as a set of recursive practices is omnipresent and never leaves you. Moreover, it seems that visible poverty is ubiquitous, and constitutes an ineluctable field. To all our shame, the sight of people having to adapt to these recursive practices in order to beg for change from passers-by remains omnipresent, too.

Regardless of local and national policies on public begging, nothing has changed in the intervening years since the team ethnography was conducted: 'If the reader hears axes being ground in the background, he has at least been given fair warning' (Watson 1970, xvi).

Leaving the field: a neglected topic

All ethnographers face the practical problem of gaining access to 'the field', whatever that might be or for how long, however straightforward or otherwise gaining access turns out to be. However, as the Introduction to this book shows, the coverage given to entering and leaving the field is heavily imbalanced. Overwhelmingly, fieldwork exits are passed over without mention.

The 'methodological appendix' to the second edition of *Street Corner Society* (Whyte 1955) has relevance to leaving the field; though this is nuanced by Whyte's own concern with whom his book would benefit, and who from the field would actually read it. These issues would be revisited in a journal special issue on *Street Corner Society*, a symposium to discuss a contentious essay (Boelen 1992) which seemed more concerned with revealing the location of the pseudonymous 'Cornerville' than addressing the methodological insights it contained. As part of his exit from the field, Whyte gave copies of his book to key informants. Whyte's anticipation that his informants would seek out the book, regardless of its 'academic' status, led to his decision to publish potentially discrediting details separately (Whyte 1943).

Although *Street Corner Society* was published by the University of Chicago Press, it is not a 'Chicago ethnography', nor of a piece with other Chicago ethnographies. Its preoccupations are found in the milieu at Harvard University, especially Whyte's mentors Conrad Arensberg and Eliot Chapple. Whyte's methodological appendix is a delight and required reading for ethnographers: it addresses crucial matters such as observer roles, prior to the publication of Gold's (1958) classic statement; and the alignments of researchers before this issue became reduced to the language game of 'positional reflexivity'. Whyte's account is fascinating also because in his 'Farewell to Cornerville' (Whyte 1955: 341 ff.) and his 'returns' to the field he realised that it had fragmented as his informants had moved on: the field no longer existed.

The furore that followed the appearance of *Small Town in Mass Society* (Vidich and Bensman 1958; see Becker 1964) could have been avoided, had Whyte's sensitivities regarding participants as potential readers of the published report been taken into account – Whyte anticipated that participants might see published research very differently than sociologists. Yet this public scandal provided an object lesson not only in regard to publication responsibilities but also in leaving the field. Abrupt or rancorous exits have consequences both for those who participated in the research – who remain 'in the field' – and also for our colleagues who follow in our footsteps. Potential informants are less likely to agree to take part in future studies if we besmirch the reputation of ethnographers through our insensitive exits from the field. Hence, there is a two-fold ethical dimension to exiting: to participants, and to future researchers who will want to gain access to fields, which may be jeopardised by our post-fieldwork actions.

For those with a research interest in public space, there are no boundary lines demarcating 'what are potential data for study' from one's everyday life. Officially entering 'the field' is contingent upon research ethics clearances, but the field is *sui generis*: it has been all around, irrespective of ethics protocols, funding and agreed work schedules. Entering and exiting 'the field' are episodic markers, treating research as linear processes rather than contingent practices. Likewise, data collection 'in the field' shades into other research activities rather than a discrete entity (Becker 1958).

Exiting the field is frequently a prosaic matter, often necessitated by the timelines of research funding, not a decision founded on a waning interest in the topic. This can be positive: characteristics of doing ethnography are that when working with people 'the more you learn, the more you see that there is to learn', and that ethnographic research 'has no logical end point' (Whyte 1955: 325). Hence, Whyte found it productive to have a shorter than expected deadline for his research because it imposed discipline on his project.

This pragmatic exit contrasts with analytic justifications for exiting the field, such as Fine's *aperçu* that within the social structure of a gaming community he had become recognised as an expert:

> with each passing week I was becoming a more and more central and powerful person within the gaming structure. I found myself unable to take a minor role, as new players increasingly came to me for advice about the game, since I was an experienced, veteran player. I found myself teaching the newer members about the game, which meant that I couldn't observe their socialization. (Fine 1983: 252)

This problematised his observational roles (Gold 1958) within this gaming community and led to his withdrawal from participation in game-play, toward emphasising different methods, and cessation of the project.

Calvey's longitudinal ethnographies of night-time economies, working covertly as a nightclub doorman, are exemplary in highlighting the problematic notion of having 'left' the field. Years after the completion of his fieldwork, he was recognised and treated as a fellow bouncer. This presented him with both social and practical problems, as he found himself re-engaging with analysis *in situ*: 'although the project was officially finished … I was never fully off "sociological duty"' (Calvey 2020: 59).

This chapter highlights an analogous case, where the fieldworker is *unable* to leave the field. However, this field is not occasioned by visits to establishments where research informants are still active. Instead, the field is *sui generis*, ubiquitous and all encompassing. The field under discussion in this chapter is public space; the phenomena are displays of visible poverty. Crucially, the phenomena of visible poverty are inextricably and reflexively linked to the social organisation of public spaces, or 'pavement cultures'.

Praxeological ethnography

An aspect of a report that was particularly valuable for our study (Lee and Watson 1993), with consequences for doing ethnographic observation, is not to see ethnomethodology and conversation analysis as related, or even as connected separates, e.g., conversation analysis merely as the ethnomethodological expression of the study of cultural methods; but to see ethnomethodology and conversation analysis as a *gestalt*, an integrated whole. Lee and Watson (1993) bring ethnomethodological and conversation-analytic insights to public space, demonstrating that, for instance, turn-taking systems are not limited to conversational environments. Adopting another insight from conversation analysis, that turn-taking is administered by the parties to the talk (Sacks, Schegloff and Jefferson 1978: 41), Lee and Watson show that when observing what happens on a pavement in public space, settings are self-administered by parties to that setting. Pavements are *self-explicating*, in that people are showing other pedestrians exactly what they are doing, such as indicating trajectories of locomotion in advance and in the process of these being transacted. This has a corollary, which is that pavements are *self-replicating*. One cohort of pedestrians will be replaced by other cohorts; and new cohorts will, in turn, exhibit self-explicating activities for each other too.

Although the study of visible poverty was an ethnographic project, it was designed to be a description of people's *practices*, and as such may be called 'praxeological ethnography'. Cathedral Square was an unfolding scene, and our original fieldwork observations take into account all the parties to the setting. This avoids reifying standard concerns, such as age, gender and race; and substantive concerns, such as categorising beggars according to technique, or comparing beggars 'and/or/vs.' pedestrians. The concatenation of identities and how these would overlap as a form of life was deliberate, an organising principle derived from a study of homeless men that was published as the inaugural issue of the journal *Ethnographic Studies* (Rose 1997).

In this chapter I discuss an encounter with a seller of *The Big Issue* magazine, and observations of begging practices. Selling *The Big Issue* enables homeless people to earn money without begging for change: as such, it may seem anachronistic to discuss this together with begging. However, discussing a seller of *The Big Issue* (hence categorisable as homeless) alongside beggars is not a conflation but a recognition that certain terms 'go together' (Harper et al. 2017). Persons standing to the side or within a pedestrian traffic flow are accountable, and, in regard to visible poverty, homelessness and begging are category predicates that 'go together'. Furthermore, there are commonalities in the use of various practices of reasoning, e.g., awareness of how membership categories are 'glance available' phenomena, members' orientations to pedestrian traffic and 'flow files', and the use of what will be described in a subsequent section as 'area knowledge'.

In a 'tutorial' situation, *The Big Issue* seller asserts that selling involves adapting to the self-organising practices of pedestrians passing by. In a later section, we shall discuss how taking Cathedral Square as a *gestalt* allowed us to look not just at those who are making themselves visible as poor but at those who are intended to see them as poor. Poverty is not necessarily automatically visible: we are attending to making it specifically noticeable, to whom and how. It is observable that beggars deploy certain strategies in order to make poverty visible, and these, like *The Big Issue* seller, are contingent upon the organisation of the street, i.e., its pavement culture.

A tutorial: selling *The Big Issue*

In this section I discuss a fieldwork encounter – securing an interview with an informant – yet this turned into a tutorial situation in which I was instructed to 'look again' at a phenomenal field. We participate in such fields every day, walking alongside other pedestrians, and the familiarity of such activities means that we take such scenes for granted. However, for some people, attending closely to pavement cultures is a way of life.

In the summer of 1997 I interviewed people who had been personally and/or professionally affected by the bombing of Manchester city centre in June 1996 (Carlin 2009). Each day that I travelled to interview appointments, walking from the platform in Victoria Station into the city centre, I saw 'Tommy' (pseudonym) selling *The Big Issue* magazine. The station had suffered structural damage in the blast wave, and fragments of the lorry containing the bomb were discovered on its roof. Trains were stopping at temporary platforms, and work routines at the station were severely disrupted. Thinking that Tommy might have been witness to the evacuation of the city the year before, I struck up a conversation with him, during which I outlined the project and asked if he would be willing to give an interview. I couldn't stand in Tommy's shoes, yet he was keen to give me some lessons in 'pavement culture' (Carlin 2017). Pavement culture turned out to be 'indigenous to, living within, and sustained by the practices' of pedestrian locomotion (Livingston 2008: 8).

One of these was at-a-glance categorisation of pedestrians. Tommy maintained that whether someone would or would not buy *The Big Issue* was seeable in advance. Tommy was not always correct, but was correct surprisingly often; what was interesting was that when he was incorrect, whether someone bought a copy unexpectedly or didn't buy a copy when he was certain a purchase would be made, he provided an account for his mischaracterisations. Tommy's success rate in anticipating buyers was not just his persistence or sheer bloody-mindedness, or a result of his engaging personality. Tommy was engaged in 'category-partitioning activities' (Lee and Watson 1993: 28), a street-level triage to maximise sales.

To explain his anticipatory prowess, Tommy encouraged me to observe closely the organisation of the pavement. As a seller of *The Big Issue*, Tommy stood in position at the entrance to the station. Yet he did not just stand still: he was *visibly* standing still. Other people stood still, but they were recognisably queuing for a ticket, or waiting for a train. Standing still is, then, an accountable matter, i.e., passers-by can determine why someone is standing still at a glance. Yet Tommy was visibly not doing waiting, nor doing queuing. Visibly standing still is consequential for Tommy, in reaching pedestrians as potential buyers. The volume of pedestrian traffic could be exploited, with occasions to stand 'within' the crowd or outside the flow, i.e., to stand on the sidelines.

A crucial aspect of Tommy's position was to show me how pedestrians themselves exploit flow files too, how pedestrians would use flow formats as cover in order to evade having to say 'No' to a seller of *The Big Issue*. It was only when Tommy put a batch of magazines in my hand in order to greet the throng of passengers disgorged from a train on the concourse – their journey between the station and the city beyond – that I realised Tommy's definition of the situation and saw the phenomenon Tommy was telling me about. Selling *The Big Issue* identifies the seller as homeless. To deny what members know as a matter of at-a-glance recognisability in the service of sociological abstraction (Carlin 2017) falsifies the phenomenon. A *moral* position vis-à-vis the seller (Becker 1967) may be taken unwittingly by adopting a particular *methodological* position, such as using carefully positioned tripod-mounted video cameras to record selling encounters that can be replayed and reviewed. This methodological option may suit the analyst but does not account for how members define their situation.

As members orienting towards an ineluctable field, one in which there is a 'standard pace' of pedestrian locomotion according to the context of one's surroundings, standing still is an incongruity, an accountable matter. In a large body of pedestrians, walking at pace from the train platform to the exit of the station, a person who is standing still within the crowd is noticeable and accountable. From the vantage point of the seller, however, pedestrians' trajectories of locomotion altered on noticing that there was a homeless person in front of them. Furthermore, these alterations, while subtle, were nevertheless perceptible.

Area knowledge

Tommy expounded in detail on the advantages and disadvantages of selling in different parts of the city centre. He outlined various criteria for what were regarded by sellers as 'desirable' pitches. Desirable pitches were places within the city centre from where the seller had access to refreshments and toilet facilities, a safe place to store quantities of the magazine in order that the seller did not have to return to the head office to collect more copies and, most importantly, footfall. The practitioners' term for such expertise is area knowledge.

'Area knowledge' (Bittner 2013) is used by incumbents of various categories such as police officers, pickpockets, beggars, street vendors, buskers, etc. Certain places attract certain crowds at certain times. The numbers of beggars on the streets increase at religious festivals, e.g., during Diwali (Patel 1959: 6), and at Easter in Jerusalem (Graham 1913: 192). Members' use of area knowledge is an awareness of the temporal and ecological ordering of the city. For the practical purposes of begging for change, members know where to go and where to stand, so as take advantage of heavier pedestrian traffic flows. Area knowledge involves the awareness not just of where to go, but when:

> Beggars have a liberal fund of knowledge about pay days. They know the factories where the workers, when they have money, are 'good'. (Anderson 1923: 43)

This ecological orientation had been noted by Friedrich Engels in the nineteenth century:

> [I]t is a striking fact that these beggars are seen almost exclusively in the working-people's districts, that it is almost exclusively the gifts of the poor from which they live. Or the family takes up its position in a busy street, and without uttering a word, lets the mere sight of its helplessness plead for it. In this case, too, they reckon upon the sympathy of the workers alone, who know from experience how it feels to be hungry, and are liable to find themselves in the same situation at any moment; for this dumb, yet most moving appeal, is met with almost solely in such streets as are frequented by working-men, and at such hours as working-men pass by; but especially on Saturday evenings ... (Engels 1986: 119)

Engels' observation highlights also that this 'particularization of knowledge' (Bittner 1967: 707) provides an awareness of when crowds of incumbents of particular categories will gather in the vicinity, providing a traffic flow of donors of higher 'potential'. Tucson, Arizona is the location of US air force and army bases, and a number of people begging on streets in Tucson are veterans of the Vietnam and Korean conflicts – current members of the defence services 'might be expected to empathize with a veteran (temporarily) down on his luck' (Williams 1995: 37). Beggars in Nigeria congregate around mosques on Fridays, the Muslim day of prayer (Bamisaiye 1974: 199).

In the following sections, the confluence of area knowledge and working the locally produced orders of pavement cultures is made visible in research on another phenomenon of visible poverty – begging.

Cathedral Square

The use of area knowledge and ecological orientations was manifest in Cathedral Square. At a large intersection in front of the courthouse building, there is a series of bus stops that are congested, due to a set of road works which have slowed the busy traffic. We arrived late morning. Apart from two women waiting at the bus stop next to us, there were not many people within view: a man in a wheelchair rolled a cigarette. One man was holding a box as he leaned against a department store window. Another man stopped and turned to face him from the other side of the pavement (a distance of perhaps six metres), standing with his back to the orange and white traffitape which sealed the road works off from pedestrians. A man with a cast on his foot limped heavily in our direction, murmuring to people as they walked past him, and stopped when he reached the pedestrian crossings behind us.

There was a sudden, remarkable transformation. At noon precisely the wide pavement filled up with people on their lunch breaks. An empty space instantly became a noisy crowd as people rushed to stand at bus stops; students stood together in small groups, becoming obstacles that the crowds had to negotiate their way around. Looking around Cathedral Square, we noticed that a number of beggars had appeared. The man with the bandaged foot was in an advantageous position, with streams of people walking past him in both directions at the change of lights. However, his main source of donations came from the assembling crowds of people waiting for the pedestrian crossing lights to turn in their favour. These crowds assembled quickly and in numbers, and each crowd was a new cohort of potential donors. The man who had been standing alone against the shop window had moved to stand next to the man in the wheelchair: he had placed the box he had been holding on a plank across the chair. The man in the chair was now furiously cranking the arm of a wind-up musical organ, twisting and turning his head to look up at individual passers-by. A different man was now in front of the department store window, attempting to catch the attention of people in the crowd as they hurried past. A young man with long hair was squatting down against the wall, playing a small pair of bongos. A man of similar age in a leather jacket silently approached people in the large queues gathering at the bus stops, before breaking into the small circles of students along the pavement.

This brief account of fieldwork observations is provided to illustrate how members use 'area knowledge': members doing begging know where to be and when. A location in the town fills up with crowds of people, coming from all directions, at noon on weekdays. Not only is this known, but it is used in practice for begging activities. Furthermore, members can position themselves at specific sites within a certain location. Area knowledge is detailed in its particulars, and can benefit its users in begging activities. In this way, poverty as visible poverty is synchronised with the daily rhythms of potential donors, who are known to be in transit through Cathedral Square at a particular time.

Preference rules

In bringing the logic of conversation analysis to public spaces, Lee and Watson (1993) observed the operation of certain 'preference rules' in the organisation of pedestrian flow files. Regularised and recurrent patterns of locomotion are demonstrated in the open public spaces at our research sites on visible poverty. Preference rules apply *mutatis mutandis* in the trajectories of flow files moving past beggars. Members in public space have a dispreference for walking between the beggar and the begging bowl.

This socially organised feature of pedestrian traffic was used as a resource by members begging on a pedestrian bridge we crossed daily, on our way from our accommodation to Cathedral Square. This bridge was frequently patronised by a man playing the harmonica. He had a terracotta bowl in front of him to collect coins. However, he had positioned the bowl in the middle of the bridge. By standing still in the same place, he remained outside the traffic flow, while his bowl was placed within it. In passing by, members manifested a preference not to walk between the man and his bowl. We did not witness a single breach of this tacit rule, indicative of its normative status (Carlin 2014: 161). At one end of the bridge, a group of beggars sat outside the flow files, on a step. However, they positioned bowls across the exit of the bridge, adjacent to the access ramp. The bridge was too narrow for pedestrians to pass by without impinging upon personal space, yet the group did not interfere with members' trajectories of locomotion. However, the confines of the entrance to the bridge ensured that, at a minimum, members had to step over the bowls: if people chose not to 'see' these beggars, they saw the bowls.

The social and visible arrangements of flow files highlighted a difference between stationary begging, whereby members doing begging are seen to be standing outside flow files and 'demonstrably "waiting on the sidelines"' (Lee and Watson 1993: 92), and practices of 'breaking' or 'interrupting' the flow momentarily in order to beg change from passers-by within the flow. This is not a comparison of, e.g., silent begging versus whispering begging, or stationary begging versus ambulatory begging. Rather, we refer to observed instances of begging activities whereby the person doing begging standing outside the traffic flow used the internal order of flow formats to engage the attention of members within the flow.

Conclusion

In developing a sociological imagination, pavements become scenes for investigating the 'sociology of everyday life', as another domain of inquiry. From this view, sociology, in its particulars, is observable by walking down the street, or sitting on a park bench (Lofland 1973). However, once the *gestalt* switch (think of bistable images such as Ludwig Wittgenstein's duck/rabbit sketch) has been made, from the sociology of everyday life to seeing pavements as perspicuous settings for the production and reproduction of ordinary methods by members for recognition by other members, 'the field' is all encompassing, omnipresent and ineluctable. There are no times out from being in the field.

Elements of transient encounters between sellers and pedestrians who purchase *The Big Issue* are available to recording technologies (Llewellyn

2011); however, video-recording should not be taken as a nostrum that solves all methodological problems. Tommy highlights a problem with video analysis, and the reification of retrievable data, by showing witnessable phenomena that are not available to recording. Analysts should be led by the phenomenon, not the contingencies of data collection. Moreover, ethnographic access to members' understandings is not possible with recorded data, which facilitate an alternate research problematic. The phenomenon should not be subjacent to documentation. In this way, praxeological ethnographies are not sociologies of a 'discoverable order' (using approved methods) but sociologies of the 'witnessable order' – 'how members of society produce and sustain the observable orderliness of their own actions' (Livingston 2008: 124). Ethnographic observation affords description of members' cultural methods from within the world in which we live, as thoroughly intersubjective matters.

What we observed in Cathedral Square was 'there for anyone to see' (MacAndrew and Garfinkel 1962); it was not unusual for the research setting, nor for other public spaces. Begging activities, constitutive of phenomenal fields of public spaces, are chronic and geographically widespread: pedestrian activities are ubiquitous in the sense that these are self-explicating and self-replicating in public spaces worldwide. While securing funding for projects to observe pavement cultures is difficult, necessary features of research, e.g., obtaining informed consent and research ethics, disguise the omnipresence of these fields, in which we are already immersed. Moreover, these are fields from which there is no discernible exit. Once you have seen that begging practices are adaptations to and exploitations of the self-explicating, self-replicating cohorts of pedestrians walking along a pavement, the field never leaves you.

Acknowledgements

With thanks to Venetia Evergeti and Ged Murtagh, without whom this chapter could not have been written; and the research directors, Rod Watson and Yves Winkin. The research was sponsored by the British Council and the Economic and Social Research Council.

References

Anderson, N. (1923). *The Hobo: The Sociology of the Homeless Man*. Chicago: University of Chicago Press.
Bamisaiye, A. (1974). 'Begging in Ibadan, Southern Nigeria'. *Human Organization*. 33(2): 197–202.

Becker, H.S. (1958). 'Problems of inference and proof in participant observation'. *American Sociological Review*. 23(6): 652–660.
Becker, H.S. (1964). 'Problems in the publication of field studies', in A.J. Vidich, J. Bensman and M.R. Stein (eds) *Reflections on Community Studies*. pp. 267–284. New York: Harper & Row.
Becker, H.S. (1967). 'Whose side are we on?' *Social Problems*. 14(3): 239–247.
Bittner, E. (1967). 'The police on Skid Row: A study of peace keeping'. *American Sociological Review*. 32(5): 699–715.
Bittner, E. (2013). 'Some elements of methodical police work'. *Ethnographic Studies*. 13: 188–194.
Boelen, W.A.M. (1992). 'Street corner society: Cornerville revisited. *Journal of Contemporary Ethnography*. 21(1): 11–51.
Calvey, D. (2020). 'Being on both sides: Covert ethnography and partisanship with bouncers in the night-time economy'. *Journal of Organizational Ethnography*. 10(1): 50–64.
Carlin, Andrew P. (2009). 'Edward Rose and linguistic ethnography: An Ethno-inquiries approach to interviewing'. *Qualitative Research*. 9(3): 331–354.
Carlin, Andrew P. (2014). 'Working the crowds. Features of street performance in public space', in T. Brabazon (ed.) *City Imaging: Regeneration, Renewal & Decay*. pp. 157–169. Dordrecht: Springer.
Carlin, Andrew P. (2017). 'Visibility and street ethnography: A lesson in pavement culture', *Etnografia e Ricerca Qualitativa*. 10(2): 287–302.
Carlin, A.P., V. Evergeti, and G.M. Murtagh (1999). 'Visible identities at work: Preliminary observations on categorisations and "street performers" in "Cathedral Square"'. Paper presented at the Annual Conference of the British Sociological Association, University of Glasgow.
Engels, F. (1986). *The Condition of the Working Class in England*. London: Grafton.
Fine, G.A. (1983). *Shared Fantasy: Role-Playing Games as Social Worlds*. Chicago: University of Chicago Press.
Gold, R.E. (1958). 'Roles in sociological field observations'. *Social Forces*. 36(3): 217–223.
Graham, S. (1913). *With the Russian Pilgrims to Jerusalem*. London: Thomas Nelson and Sons.
Harper, R., Watson, R. and Woelfer, J.P. (2017). 'The Skype paradox: Homelessness and selective intimacy in the use of communications technology'. *Pragmatics*. 27(3): 447–474.
Lee, J.R.E. and Watson, D.R. (1993). 'Public Space as an Interactional Order'. (Final Report to the Plain Urbain). Unpublished manuscript, Department of Sociology, University of Manchester.
Livingston, E. (2008). *Ethnographies of Reason*. Aldershot: Ashgate.
Llewellyn, N. (2011). 'The delicacy of the gift: Passing donations and leaving change'. *Discourse & Society*. 22(2): 155–174.
Lofland, Lyn H. (1973). *A World of Strangers*. Prospect Heights, IL: Waverley Press.
MacAndrew, C. and Garfinkel, H. (1962). 'A consideration of changes attributed to intoxication as common-sense reasons for getting drunk'. *Quarterly Journal of Studies on Alcohol*. 23(2): 252–266.

Patel, T. (1959). 'Some reflections on the beggar problem in Ahmedabad'. *Sociology Bulletin*. 8(1): 5–15.
Rawls, A.W. (2009). 'Communities of practice vs. traditional communities: The state of sociology in a context of globalization', in G. Cooper, A. King and R. Rettie (eds) *Sociological Objects: Reconfigurations of Social Theory*. pp. 81–99. Farnham: Ashgate.
Rose, E. (1997). 'The unattached society'. *Ethnographic Studies*. 1: xv–43.
Sacks, H., Schegloff, E.A. and Jefferson, G. (1978). 'A simplest systematics for the organization of turn-taking for conversation', in J. Schenkein (ed.) *Studies in the Organization of Conversational Interaction*. pp. 7–55. New York: Academic Press.
Vidich, A.J. and Bensman, J. (1958). *Small Town in Mass Society*. Garden City: Anchor.
Watson, G. (1970). *Passing for White*. London: Tavistock.
Whyte, W.F. (1943). 'A slum sex code'. *American Journal of Sociology*. 49(1): 24–31.
Whyte, W.F. (1955). *Street Corner Society*. (2nd edn). Chicago: University of Chicago Press.
Williams, B.F. (1995). 'The public I/eye: Conducting fieldwork to do homework on homelessness and begging in two US cities'. *Current Anthropology*. 36(1): 25–51.

10

'The martial will never leave your bones': embodying the field of the Kung Fu family

George Jennings

Embodiment and apprenticeship in martial arts studies

As Wile (2015) has remarked, just as women have developed women's studies and LGBTIQ people have created queer studies, martial artists have created martial arts studies as a labour of love. Indeed, martial arts scholarship, now recognised as a distinct field within English-language academia as 'martial arts studies' (see Bowman 2015), is almost the exclusive product of practitioner-researchers or 'pracademics' who are long-term advocates, students and even instructors of a given art. Aikido practitioners usually research Aikido, and Capoeiristas are quite often those who write about Capoeira. Many of these scholars were practitioners before they became researchers, while others became practitioners through the position of an apprentice of similar mind–body practices (Downey, Dalidowicz and Mason 2014). Indeed, many ethnographers continue to practise what they initially learned just for research purposes, taking the bodily movements and methods in their daily lifestyles. As I have noted, it is the body that has received a great deal of attention, as seen in decades of scholarship pertaining to embodied experiences, interactions, body pedagogy and violence (Channon and Jennings 2014), as well as the numerous monographs centred on martial arts bodies (Jennings 2018a).

Logically, then, research on Kung Fu – the popular, generic term for the Chinese martial arts – is no exception, as it is studied from the standpoint of the practitioner-researcher interested in embodiment of skill and culture. The different renderings of Kung Fu, Gung Fu or Gong Fu are relatively modern terms with recent connotations with the martial arts. Kung Fu literally means 'hard work achieved through time and effort', and some (such as my Taijiquan instructor's own teacher) even translate this as 'a marriage to the pursuit of skill', as the Gong pertains to skill and the Fu can be understood as a lifelong commitment or marriage. Kung Fu is, as my fellow practitioner-researcher Veronika Partiková and I have found (Partiková and Jennings

2019), all about relationship and family – relationships between teacher and student, *Kung Fu* 'brothers' and 'sisters' (in the fraternal sense) and a sense of place or home in the school (Kwoon). The relationship even extends to one's Kung Fu ancestors – those founders and innovators along one's martial arts 'family tree'. The southern Chinese martial arts that we and other scholars such as Judkins and Nielson (2016) have studied are historically categorised into different regions and 'families'. Many of these were indeed, at some point, the product of specific clans who dominated towns and villages, and they still retain the names of these families, as in those systems finishing in the Cantonese term Gar (Chow Gar, Mok Gar, etc.) and Kuen (fist). However, in today's globalised, mobile, multicultural world of Kung Fu, in which a White English person like me might receive a black sash (itself a recent invention) from his White English teacher, the family is largely an intergenerational, intercontinental and pluri-ethnic connection between practitioners who come from the same lineage of teachers – tracing their Kung Fu ancestry to a key figure or semi-mythical founder who made significant technical contributions.

These lineages are therefore not normally biological lineages or family trees but, following Lakoff and Johnson (1980), metaphorical ones. They are in fact what David Brown and I (Brown and Jennings 2011) have called 'body lineages', as they are concerned with the transmission of embodied skill from one generation to the next in order to keep the art of style in question alive. This embodied skill, which is the result of painstaking effort, must be replicated into the bodies and movements of normally younger practitioners who are expected in turn to continue passing down this embodied knowledge as part of a specific, institutional martial habitus: durable schemes of dispositions that eventually become subconscious after conscious training. This connects with the recent theoretical ideas of Spatz (2015), who postulates that: (1) technique is a form of knowledge, and (2) the practice of technique is a form of research. He also states that, once learned, technique is a form of knowledge that is not easily unlearned. In Kung Fu, notable techniques might include ways of standing, footwork movements, punches and kicks, blocks and controls. When repeated thousands of times, they mould the body into a certain shape and capacity that remains inside the practitioner for years – or even a lifetime.

Meanwhile, one of the most prominent ways of researching these techniques is ethnography, as seen in the numerous collections on martial habitus and techniques of the body within specific case-study schools (Sánchez-García and Spencer 2013; Nardini and Scandurini, forthcoming). These follow pioneering ethnographies such as those of Wacquant (2004) and Zarrilli (1998), who, as devoted apprentices, became practitioner-researchers of boxing and the lesser-known art of Kalarippayattu, dutifully learning under the watchful gaze of their coach and *gurukkal* (a guru who is part of a long line of gurus). The notion of habitus is central to many ethnographies, with certain scholars and entire monographs advocating phenomenological approaches to study the sensuous aspect of martial arts practice (Spencer 2013), figurational perspectives on the civilising and decivilising processes connected to martial habitus (Ryan 2016; Sánchez-García 2019) and ideas around diaspora and institutional habitus as inspired by Bourdieu (Delamont, Stephens and Campos 2017) – all of which shed light on the incredible variance between individual practitioners and institutions in terms of how technique is conceptualised, trained and performed in lineages and academies.

Nevertheless, it is important to point out that ideas of pure lineages and families simplify the realities of cross-training and movements between and within styles that has occurred for a long time in the Chinese martial arts. Moreover, today's *Sifus* (teachers; literally 'teaching father') of Kung Fu are likely to have trained in numerous martial arts from their youth, and may learn Taijiquan (Tai Chi Chuan) in their later years and perhaps other arts such as massage or Qigong (Chi Kung) from different masters. Some even use Western training methods such as weights and kettlebells as well as sparring to develop their bodies and combative abilities while testing out new techniques. This combination of body practices and cultures may stimulate the innovation of a new approach to Kung Fu that is more concerned with posture and well-being or even spirituality. Such new derivatives of Kung Fu are the product of specific dispositions forming part of the martial habitus: the specific schemes of dispositions for combat efficiency, technical transmission and intellectual, spiritual and ecological understandings of the art in question (Brown and Jennings 2013).

Wing Chun is quite possibly the most widely practised form of Chinese Kung Fu, largely owing to the fact that it was the martial art of the young Bruce Lee (1940–73), who learned from Grandmaster Ip Man (1893–1972). I am a student of this most famous branch of Wing Chun, which is now mythologised in the various Hong Kong biopics such as the *Ip Man* films (2008–19) and the Academy Award-nominated movie, *The Grandmaster* (2013). The system is compact (normally composed of three empty-hand forms, a wooden dummy form and two weapons sequences), and it is mainly taught through a cooperative drills-based pedagogy concerned with close-quarters fighting for self-defence. The art has diversified over the last few decades after the death of Ip Man, with revelations of other branches emerging as well as numerous innovations from newer generations of practitioners – most notably in drills and an investigation into the softer, 'internal' aspects of the art. These continued technical and pedagogical developments and historical revelations make Wing Chun an interesting art to study as a social scientist. Although Wing Chun Kung Fu is my main martial art and is central to my identity as a martial artist, I had prior experience of Taekwondo and Kendo, and later Judo, before continuing my research into the newly founded Mexican martial art of Xilam (see Jennings 2015; 2016; 2018b). I am now investigating both historical European martial arts (HEMA) and Taijiquan in terms of their linguistic pedagogies through a dual, contrastive ethnographic project.

As a teenager, I initially learned Wing Chun for three years in my home town after being attracted to the art for its directness and simplicity after spending time learning the more athletically demanding Taekwondo and the more ritualistic Kendo. The congruence between the solo forms, partner drills and application drew me to the art, as did the potential longevity, as seen in a noted grandmaster still training and teaching in his late seventies. Later, as a university student in another region of England, I had to relearn how to stand, turn, step and hold my techniques as I went through seven continuous years of being a student in the Academy under the direct tutelage of Bridge and his senior students. This initial reconstruction of my initial habitus developed a very specific martial habitus of 'Bridge's Wing Chun', which is based on the teachings of his four main teachers, giving it distinct ways of conducting Wing Chun techniques and shapes, such as the stance, blocks and angular punches. To take the popular tree metaphor, the international family of Wing Chun has over seven major branches which themselves are divided into many associations and federations. For reasons of the politics of global Wing Chun and its many feuds, all names are pseudonyms beyond this point. In what follows, I present brief periods of my life as a practitioner, practitioner-researcher and ethnographer of the martial arts in terms of my gradual embodiment and adapted disembodiment of a specific Wing Chun sub-field.

A practitioner-researcher's tale

In this confessional tale, I reveal insights from how this system was so embedded into my body that it transcended conscious thought. The Wing Chun techniques trained over the seven years in Bridge's Academy were, quite literally, in my flesh and bones, and could not be removed without years of conscious effort – as I had done when I first switched lineages. In what follows, I argue that embodied technique is the basis of what the Kung Fu family develops and, as such, that one does not easily leave this family – as it is inside the practitioner-researcher. As my first Taijiquan teacher once commented on my Wing Chun-like movements in his more flowing art, 'the martial will never leave your bones, will it?!' This comment was actually a rather negative remark, as he was referring to the tension in my limbs, which needed to be relaxed along with my mind. In the following four sections, I share insights into my experiences of: (a) becoming a practitioner-researcher; (b) leaving the proximity of the Kung Fu family; (c) how I found the art of Wing Chun in different areas of life; and (d) the ways in which I have later attempted to research other martial arts by attempting to override my instincts as a practitioner.

Researching embodied pedagogy

None of my real biological family has ever practised martial arts or combat sports, so I remain the only member who has studied different forms of fighting. I began martial arts training in 1998 at the age of 14 with Taekwondo, as my best friend, Tim, was and still is an exponent of the art. A year later, I found Wing Chun Kung Fu at 15, after reading a biography of Bruce Lee, *Fighting Spirit* (Thomas 1994), which was a Christmas present from my mother, who soon recognised that my obsession with the martial arts had replaced that of Star Wars. The directness, simplicity and compact nature of Wing Chun immediately appealed to my personal logic. After three years in my local Kung Fu association, I moved to study at university, and was keen to find a Wing Chun school to continue training with. Little did I know how different this style would be – with its more direct approach to fighting, limb conditioning and vast numbers of partner drills. I was so impressed by this approach to Wing Chun that I decided not to return to my first school. After being inspired by Zarrilli's (1998) impressive monograph of the Indian martial art of Kalarippayattu, *When the Body Becomes All Eyes*, at the end of my second year at university, I aspired to become a martial arts scholar. My immediate family were doubtful that this was possible, but this was understandable, as none of them knew much about academia. Notwithstanding the fact that 'martial arts studies' as it is now called was not so recognised then, I was determined to conduct an ethnography for my undergraduate dissertation from 2004 to 2005. This continued with my master's dissertation (2005–6) and then a life-history study for my PhD (2006–10), all at the same university close to Bridge's Martial Arts Academy and its two branches in two different towns. As I spent more and more time in the university with postgraduate studies, I had less time with my real family, and more hours honing my skills with my adoptive Kung Fu family.

Continuing with the family and home metaphors, many Chinese martial arts schools have the idea of 'indoor students' or 'closed-door students' who receive a deeper level of tuition in order to receive the entire martial art with all its intricacies. These indoor students are selected after years of loyal study with their teacher in the manner of a disciple. My own *Sifu* explained this concept to some of us long-term (and long-distance) students during the COVID-19 lockdown through online lectures to his followers. According to him, the majority of students were 'skin and hair' – they did not have the correct attitude and diligence to receive the full art, so were given the basics of the art in return (the skin and hair). But there were a small number of talented and caring devotees who were supposed to receive the deeper aspects of the art, such as its theories and more hidden martial applications. These received the 'bones and muscle' of the art. For an ethnographer like me, the art would be absorbed into my flesh (as in muscle fibre types, callusing and muscle memory) and bones (bone density through conditioning, and even posture).

My muscles and bones had to change between two styles of the martial art. The first school of Wing Chun was in my home town, and it taught a 70/30 weight distribution, with the body mass being rested onto the back leg for a more defensive strategy. The second school, Bridge's Academy, is its own branch of Li's Wing Chun (Master Sebastian Li being Bridge's teacher), which follows a more unusual policy of a balanced, 50/50 weight distribution accompanied with a more aggressive approach involving the conditioning of limbs, which hardened my bones and raised my pain tolerance. Li has studied Monkey Boxing (an athletic, animal-inspired form of Kung Fu), Taijiquan, the preparation of Dit Dar Jow oil and related deep-tissue massage, as well as lion dancing, while his position as a challenge fighter in his younger years for his own teacher, the late Grandmaster Long, provided him with plenty of combat experience. As such, Li's system boasts a variety of angles for attack, and keeps a tight geometric alignment of the limbs in order to absorb incoming force. Due to these considerable differences between the styles, I opted for the second school as my new Kung Fu home, Church Kwoon, and did not return to my old home town school in order to avert political tensions between the groups and to avoid technical confusion. I then researched this school from 2004 to 2009 as part of an ethnographic and life-history study (see Jennings, Brown and Sparkes 2010).

Leaving the physical presence of the family

As my PhD came to an end, I moved to London, and later Scotland, to pursue my academic career as a lecturer. 'It's not like you're going to Russia!' said one of my Kung Fu brothers as I looked doubtful that we would see each other very often. However, I did move further afield: across the Atlantic. Some years later, while living in Mexico (2011–16), I took up the offer of a free diagnosis of my alignments and weight distribution with a chiropractor. This was following a time when I had helped a sport science colleague with her research project in London. She had told me, to my surprise: 'I've never known someone so over-developed on one side. Is your sport very one-sided?!'

Wing Chun is actually a symmetrical system in principle, as each movement is practised on the left and right hand in the form sequences in order to develop ambidextrous fighters. But, in normal classes, the system is adapted for people living in a predominantly right-handed world. Hence my right scapula was folded forwards, leading to an early stage of sclerosis as identified by the chiropractor. This added to the more obvious rounded shoulders and upper back that other people had warned me about, which was due again to an over-emphasis on forward, explosive movements in contemporary Ip Man Wing Chun – famous for its aggressive, rapid-fire punching. The cost of the chiropractor was significant for a university and English teacher in Mexico City, so from then on I spent more time on my standing postures in an effort to heal myself to some degree. This led me to begin investigating the potential of stance training in the Chinese martial arts from an auto-phenomenographical perspective (see Allen Collinson, Vaittinen, Jennings and Owton 2018). This entailed studying my own embodied sensations through progressively longer periods of standing in the basic training stance of Wing Chun in order to determine the role of the sense of heat (thermoception). I collaborated with my three colleagues in the UK, who are practitioner-researchers of distance running, mixed martial arts and boxing.

Shortly after initiating that collaborative autoethnographic study, one of my co-authors, Anu Vaittinen, started to study the pedagogy of Wing Chun in our own small groups (Jennings and Vaittinen 2016). She was a student of her life partner, who had a small, informal Wing Chun school in the north-east of England. I was actively teaching Wing Chun in my old university in Mexico, Universidad YMCA, with a handful of university students and older members of the public. We interviewed the students about how they prepared for the classes and what they did to supplement their learning. We found that the students of different generations and cultures were all actively using digital technology and internet resources such as YouTube to assist their understanding of technique, practise at home and motivate themselves for the class. All of this preparation was akin to what I used to develop my own skills: watching films and documentaries about Wing Chun and Kung Fu, training with instructional videotapes and reading intensely – albeit in a less digital fashion. This had helped to imprint the specific Wing Chun family into my DNA, despite the fact that I was halfway across the world from my principal *Sifu*.

Living embodied skill

Thanks to years of training and teaching Wing Chun, the shape of techniques in movement has become automatic to me, and this has become apparent when I have been in predicaments in 'the street'. Fortunately, I have not had to use my Wing Chun skills in a survival situation, but I have found myself utilising the techniques without thinking about them. During my time in Mexico City, I was often approached by people due to my different physical appearance. Due to the tactile and friendly nature of many of the people in this 'touchy' Latin culture coming too close for comfort, I used techniques to ward off unexpected contact. For instance, while running late for a business English class with a burger in my hand, a strange man moved from his street stall to touch me. 'Guëro!' he called in an extravagant manner, using the Mexican vernacular term for blonde or fair-skinned people. Without thinking about it, I continued to walk forwards and raised my left guard hand to the Wu Sau position, first forwards to the centre, and then drawing back on the elbow, which sent the man spinning to the side, collapsed onto his stall. Now safe, I continued in motion to my class, amazed by what had happened: I had defended myself without thinking!

On a later occasion, I was walking back from a class in a different company. To my right, I could see some builders' labourers having a play fight in the street, so I moved to the left side of the pavement to give them more room. I had seen such men fight before, and knew they could be dangerous adversaries. One of the men gave his workmate an almighty shove, making him stumble several metres back, and the man was about to crash into me, despite my recalculations of a safe distance. Without any time to think, my body did the work for me: I turned 90 degrees to my right with a Lan Sau (bar arm) technique, which absorbed the body weight and force of the builder, who bounced off me as I automatically turned off the technique, continuing my walk as if nothing had happened.

Finally, while waiting outside a building near my old home one dark evening, I was approached by someone from my left. Immediately, I brought my triangular guard upwards and forwards to present this between me and the man. It was the friendly vendor of *esquites* (a popular form of Mexican street food) who sold me this snack on a regular basis. 'Sorry!' I called. Despite my year of learning the Mexican martial art of Xilam and several years away from my Kung Fu family, the movements were an automatic part of my habitus. But could something be done to change this? These reflexes were definitely pleasing and useful for self-defence situations, but they could hold me back when learning a new martial art for my research career.

Changing technique and habitus

My current Taijiquan and internal martial arts teacher, David, has now welcomed me to his group of 'indoor students', once joking that I must be an indoor student as the stiff door to the hall we train in once opened for me. Yet it took me a similar period of a year to move in a more Tai Chi-like fashion. I could recall and replicate the movement sequences of specific segments of the Yang Short Form quite well, but not with the correct principles expected. 'It's...very Wing Chunny', David told me with a slight smile after I demonstrated what I had been practising in my corner of the hall. I was expected to move from the centre rather than drive my movements from the hands. Again, the martial aspect of Wing Chun was retained in my bones.

Was the Kung Fu I had learned superficial, outdoor student material? In recent years, Master Li has been revealing more of the deeper, 'internal' aspects of the art, and our own *Sifu* is doing the same. We are being taught to move from the centre, and are spending more time discussing concepts similar to those found in Taijiquan. As such, the techniques of the system had been refined to become more economical and softer upon contact. After my autophenomenographic study of the standing postures, this was already apparent to me. Upon my return to the UK in 2016, I was able to reunite with my *Sifu* and the wider Kung Fu family through private lessons with *Sifu* and seminars. Meanwhile, I train on a weekly basis with a former student of *Sifu*'s (incidentally, also called George and an English 'ex-pat' living in Wales). As a consequence, I feel a more central part of this association and have maintained my embodied reflexes, although I continue to work on my posture and correct alignments. During the COVID-19 lockdown, crisis turned into a moment of creativity for many martial arts groups, including Bridge's Wing Chun Academy. *Sifu* offered online classes and lectures on Wing Chun using the Messenger application, as he keeps an active Facebook group in which we could watch short videos, observe archive photographs and discussion posts for later shared analysis. This forms part of an ongoing study as my ethnographic work turns to pragmatism concerned with changing body pedagogics (Jennings, 2020).

My body has of course changed over the years of apprenticeship in Wing Chun and other martial arts. I am more relaxed, thanks to Taijiquan, and understand more about armed combat, due to HEMA. Fifteen years later, however, I can safely say that my old Taijiquan teacher was right with his 'The martial will never leave your bones, will it?' The embodied skill and alignment of Wing Chun is still deep within me as I am trying to unlearn some of its principles in order to study other martial arts. The martial – in this case the Wing Chun – has not left my bones. Even now, in my current fieldwork on HEMA, my fencing instructor had to spend a good year correcting my leg alignment, as it was facing in at 45 degrees – the exact position for my knee to protect me from a groin kick. But when fighting an opponent with a sword held in a central guard, a kick to the groin would be a foolish move. Despite my understanding of the different logic of medieval sword-play, my leg just would not turn straight for the toes and knee to face my training partners. The website image of the Blade Academy revealed this. Under our black masks and armour, I could instantly recognise myself in an action shot along the Viking shield wall from the position of my foot. From the waist up, I looked like a HEMA enthusiast or a battle re-enactor, but from the hips down, I was still a Wing Chun man. Perhaps I had taken the field of Wing Chun and installed it within me.

Conclusions and implications: when a scholar embodies a field

Martial arts studies is a field founded and developed by practitioner-researchers, and ethnography remains a key tradition to study the broad theme of embodiment: the living, feeling, moving body that form the basis of such body cultures (Eichberg 1998). Indeed, this chapter is one of two contributions in this book on a martial arts practitioner-researcher's apprenticeship and journey (see Chapter 14). Recently, Paul Bowman, one of the pioneers of martial arts studies, has asked scholars about what we embody as practitioner-researchers (Bowman 2019). Different practices offer a different form of embodiment – from embodying strength in weight training to ideas of flexibility in yoga. Defensive strategies? Postural alignment? Reflexes for self-defence situations? The martial arts are a good example of bodily practices in which a field can be embodied for many years – even a lifetime. This might hold true for other mind–body disciplines and the specific changes in particular body parts.

Future research could draw on ideas on particular body parts, as in Bates' (2018) study of people living with long-term illnesses. In a wider context, transformation of the body might include the flexibility of the pianist's fingers, the dancer's sense of rhythm or a yogi's sense of balance. I hope this confessional tale is useful in some way to stimulate specific considerations of how we do not just study embodied fields such as the martial arts but actively embody them through and beyond our fieldwork. Following Spatz's (2015) insights into the difficulty of losing embodied knowledge, such considerations might consider not just how we acquire skill through our apprenticeships within a fieldwork site, but how we modify this skill upon leaving the field, and perhaps lose it to some degree. In other cases, such as mine, the skill is still deeply entrenched within the body. For me, although I have left the fieldwork site, it is within me: my identity, the way I move and stand, fight and learn new things. It has, quite literally, yet to leave my bones. Friendship, and interests in self-defence and discipline, first introduced me to the martial arts, but notions of fictive kinship have kept me in a specific, localised branch of the global Wing Chun family: one that launched my career as a 'fighting scholar' and one that now permits me to continue to learn at a steady pace as part of potentially lifelong cultivation project for health, embodied knowledge, longevity and shared discovery.

References

Allen Collinson, J., Vaittinen, A., Jennings, G. and Owton, H. (2018). 'Exploring lived heat, "temperature work" and embodiment: Novel auto/ethnographic insights from physical culture'. *Journal of Contemporary Ethnography*, 47(3): 283–305.

Bates, C. (2018). *Vital Bodies: Living with Illness*. Bristol: Policy Press.

Bowman, P. (2015). *Martial Arts Studies: Disrupting Disciplinary Boundaries*. London: Rowman & Littlefield.

Bowman, P. (2019). *Deconstructing Martial Arts*. Cardiff: Cardiff University Press.

Brown, D. and Jennings, G. (2011). 'Body lineage: Conceptualizing the transmission of traditional Asian martial arts (in the West)'. *Staps*, 32(93): 61–71.

Brown, D. and Jennings, G. (2013). 'In search of the martial habitus: Identifying dispositional schemes in Wing Chun and Taijiquan', in R. Garcia Sánchez and D. Spencer (eds) *Fighting Scholars: Habitus and Ethnographies of Martial Arts and Combat Sports*. pp. 33–48. London: Anthem Press.

Channon, A. and Jennings, G. (2014). 'Exploring embodiment through martial arts and combat sports: A review of empirical research'. *Sport in Society*. 17(6): 773–789.

Delamont, S., Stephens, N. and Campos, C. (2017). *Embodying Brazil: An Ethnography of Diasporic Capoeira*. London: Routledge.

Downey, G., Dalidowicz, M. and Mason, P.H. (2014). 'Apprenticeship as method: Embodied learning in ethnographic practice'. *Qualitative Research*. 15(2): 183–200.

Eichberg, H. (1998). *Body Cultures: Essays on Sport, Space and Identity*. London: Routledge.

Jennings, G. (2015). 'Mexican female warrior: The case of Marisela Ugalde, the founder of Xilam', in C.R. Christopher and A. Channon (eds), *Global Perspectives on Women in Combat Sports*. pp. 119–134. London: Palgrave Macmillan.

Jennings, G. 2016. 'Ancient wisdom, modern warriors; The (re)Invention of a Mesoamerican warrior tradition in Xilam'. *Martial Arts Studies*, 2: 59–70.

Jennings, G. (2018a). 'Ethnography in martial arts studies', in P. Atkinson, S. Delamont and M. Williams (eds) *The SAGE Encyclopaedia of Social Science Research Methods* (online handbook). Los Angeles, CA: SAGE. https://methods.sagepub.com/foundations/ethnography-in-martial-arts-studies (accessed 25 January 2023).

Jennings, G. (2018b). 'From the calendar to the flesh: Movement, space and identity in a Mexican body culture'. *Societies*. 8(3): 66.

Jennings, G. (2020). 'Martial arts under the COVID-19 lockdown: The pragmatics of creative pedagogy'. *Sociología del Deporte*. 1(2): 13–24.

Jennings, G. and Vaittinen, A. (2016). 'Multimedia Wing Chun: Learning and practice in the age of YouTube'. Kung Fu Tea blog: https://chinesemartialstudies.com/2016/09/08/multimedia-wing-chun-learning-and-practice-in-the-age-of-youtube/

Jennings, G., Brown, D. and Sparkes, A.C. (2010). '"It can be a religion if you want": Wing Chun Kung Fu as a secular religion'. *Ethnography.* 11(4): 533–557.

Judkins, B. and Nielson, J. (2016). *The Creation of Wing Chun: A Social History of the Southern Chinese Martial Arts.* Albany, NY: State University of New York Press.

Lakoff, G. and Johnson, M. (1980). *Metaphors We Live By.* Chicago, IL: University of Chicago Press.

Nardini, D. and Scandurini, G. (eds) (Forthcoming). Special issue: Hand-to-hand sports and the struggle for belonging. *Ethnography.*

Partiková, V. and Jennings, G. (2019). 'The Kung Fu family: A metaphor of belonging across time and space'. *Revista de Artes Marciales Asiáticas.* 13(1): 35–52.

Ryan, M.J. (2016). *Venezuelan Stick Fighting: The Civilizing Process in Martial Arts.* London: Lexington Books.

Sánchez-García, R. (2019). *The Historical Sociology of Japanese Martial Arts.* Abingdon: Routledge.

Sánchez-García, R. and Spencer, D.C. (eds) (2013). *Fighting Scholars: Habitus and Ethnographies of Martial Arts and Combat Sports.* London: Anthem Press.

Spatz, B. (2015). *What a Body Can Do: Technique as Knowledge, Practice as Research.* London: Routledge.

Spencer, D.C. (2013). *Ultimate Fighting and Embodiment: Violence, Gender and Mixed Martial Arts.* London: Routledge.

Thomas, B. (1994). *Bruce Lee: Fighting Spirit.* London: Pan.

Wacquant, L.J. (2004). *Body and Soul: Notebooks of an Apprentice Boxer.* Oxford: Oxford University Press.

Wile, D. (2015). 'Book review: The creation of Wing Chun'. *Martial Arts Studies.* 1: 83–85.

Zarrilli, P.B. (1998). *When the Body Becomes All Eyes: Paradigms, Discourses and Practices of Power in Kalarippayattu, a South Indian Martial Art.* New Delhi: Oxford University Press.

Part III

Intermissions and returns

11

Between open and closed: recursive exits and returns to the fuzzy field of a community library across a decade of austerity

Alice Corble

Introduction

This chapter explores lessons that can be learned from engaging with an ethnographic field site through a series of recursive exits and returns, produced through an entangled relationality to the 'fuzzy' field of library practice through multiple and intersecting subject positions of researcher, volunteer, employee and activist. I survey how both the field itself and my compound positionality and identity within it changed over time, focusing on productive tensions arising from the difficulties of entanglement and extraction. In exploring how difficult it was for me to leave the field through a series of returns, I consider the important role of volition and affective ties in my oblique relationality to the field. I argue that methodologically messy and longitudinal back-and-forth movements between leaving and returning, theory and practice, provide a challenging yet powerful analytic opportunity for enriching what 'the field' and one's position in, around and through it can mean in both ethnographic discourse and activist practice.

The empirical field site in question is a volunteer-run library called New Cross Learning, where I was involved between 2010 and 2020. New Cross, situated in the south-east London Borough of Lewisham, is a densely populated, diverse area, with significant socio-economic deprivation in the neighbourhoods close to the library. Following the first wave of austerity public spending cuts in 2010, Lewisham borough council withdrew five of its thirteen libraries from public service delivery and funding. Four of these were transferred to voluntary sector organisations via a process of 'community asset transfer', but New Cross Library was not offered to any bidders and was closed, prompting a lively campaign to save it. A local government commentator reported Lewisham as being 'one of the first major councils to experiment with Big Society-style cultural services provision' (Conrad 2010). In spring 2011 the library was reclaimed by a group of local resident activists who formed themselves into a 'New Cross People's Library' committee. This was the result of a series of high-profile campaigns and direct-action protests, including an overnight occupation carried out in conjunction with students and activists who coalesced around concurrent anti-austerity struggles at the time. As a local resident in the borough, I witnessed these actions and participated in some of the campaign events and meetings.

The present chapter charts my ethnographic and activistic entanglements with New Cross Learning across the past decade, describing my experience in specific time periods that saw my repeated entries and returns, before discussing and analysing the significance of these entanglements and extractions. The chapter concludes with three methodological lessons to be learned from this praxis-oriented journey.

2013–14 Entry and exit

When I entered New Cross Learning in summer 2013 to begin my year-long ethnography there I was not a stranger to the library. It had been easy to gain access, as I was known to the lead volunteers, Kathy and Gill, as a local resident and via my involvement in the campaign to save the library back in 2010–11. Although this previous involvement had been a peripheral one, it became the catalyst for my sociology PhD on the civic politics of public libraries in times of crisis, which began at the nearby Goldsmiths University in 2012. My overarching research question asked how English public libraries can maintain their foundational identity as agents of social change, while being simultaneously forced to radically change their organisational forms and capacities in a crisis context. The previous two years of the newly formed 'People's Library' had been full of excitement and somewhat anarchic. By mid-2013 its organisational structure became more formalised via partnerships with a local neighbourhood charity and Lewisham borough library service, which still provided the books but no funding. At this point the library also changed its name to New Cross Learning (NXL).

My entry to NXL in October 2013 was as a familiar stranger, with the newly acquired identity of ethnographic researcher. Given my previous professional experience in both public and academic libraries in other London boroughs, Kathy and Gill were keen to put me to work. It was agreed that I would begin working as a regular part-time volunteer as a way of ethnographically observing and participating in the library's everyday life. The majority of NXL volunteers had never worked in a library before. My relative ease in helping to run the basic operations of the library meant that relationships of mutual trust and learning with other volunteers and users developed quickly. I was careful not to position myself as an expert in any way, which was not difficult since working at NXL was not like any other library I had known and there was always something new to learn.

Sometimes in the field I couldn't be an ethnographer, as I was too busy being a worker. The library largely ran on circuits of goodwill and care, with not much of a system of managing or predicting who would turn up for volunteer shifts each day. Kathy and Gill were there every day without fail as the managing volunteers, but also often juggled their work running, cleaning and caring for the library and its often-vulnerable users with caring for their young grandchildren in the space at the same time. In order to give them a break, I would sometimes be a lone worker fully absorbed in the multiple tasks at hand. I would frequently go home too tired and too much in 'worker mode' to know what to write in my field diary, and my ethnographic mind would feel overloaded or confused.

In addition to helping with the core work of running the library in terms of supporting books and information usage, I became involved in organising and participating in a range of public events and group meetings that took place in the library, such as film screenings, talks and workshops by political activists, artists, authors and community organisers. The library became an organising hub for local campaigns, such as stopping the borough's hospital accident and emergency department from closure. There were some extraordinary public events during this time, and always with a social justice focus. There was a talk given by two original Black Panthers on their revolutionary struggle, as well as a film screening, talk and Q&A with the socialist filmmaker Ken Loach. Kathy and Gill also made use of my time and skills in providing administrative support for their extra-curricular work as activist members of the local Labour Party, helping to type up documents and circulate campaign publicity via the networks that circulated in and around the library. This was a distinct change from my previous experience and understanding of public librarianship, which was grounded in professional ethics and organisational policies of political neutrality.

Nine months into my time at NXL, I conducted twelve in-depth, semi-structured interviews with library volunteers, users and associated professional library staff who worked for the local authority. I based my questions along temporal lines, asking participants how their experiences of NXL had impacted on or changed their past, present and future conceptions of library spaces and services. The following interview excerpts illustrate a key theme of 'openness', revealing the radical potentiality of what a library can do and be when it is not governed by the state but, rather, placed in the hands of its people:

> I feel more comfortable here than in your average library because it hasn't got the sanction of official space – I think it's the unofficial anarchic feel I like

about it. The fact that everybody's here because they want to be and not because they've got a career as a librarian – they're ordinary citizens like yourself. We're all sticking together to reassert ourselves as individuals in a growingly oppressive social state. Spaces like this are essential for your ordinary individual to come to exchange like-minded views with his fellow citizens. […] You can feel that you're in an open space. Wide open – embracing – free – wide open. When it's open it's really open. (Jackson, regular NXL library user 2014)

An interview with another participant reinforced this, as she talked about volunteers' capacity to be open-minded and open-hearted, which was part of how the library operated 'along the lines of open sourcing – the people who use it are part of what it is, part of the process of evolving' (Jennifer, NXL volunteer 2014).

Once the NXL interviews were complete in late summer 2014, I wound down my volunteer library shifts and followed my research plan of exiting the field at this point, as I was soon to move on to ethnographic research in another library field site. And so came my first ethnographic exit from NXL, which at this point was thought to be the only departure; however, there were several returns that would bring me back to this site and complexify my entanglement with it.

2016–17 Re-entry and exit

Although I did indeed exit the NXL field in 2014, the subsequent research did not go according to plan, as personal crises intervened and I needed to take a break from the PhD and step back from research. However, I did not leave the broader 'field' at this juncture, as I remained in the library sector professionally as an agency worker in academic and public libraries, which included six months as a data-driven librarian for Lewisham council, employed to analyse the ways in which the borough library service demonstrates value for its citizens. This put me back into proximity with NXL, but at the same time at a distance, a few miles away in the offices of the council headquarters. At this point the library service had recently incurred a second round of cuts with a further £1,000,000 removed from its budget, five more libraries outsourced to volunteer management and around 100 library staff made redundant. In such a decimated service context, it was virtually impossible for me to 'demonstrate value' of the library service as my job description prescribed, particularly as the measures used by the council were purely quantitative based on library book borrowing and footfall. I left the job disillusioned, but with a renewed sense of the value of ethnography in capturing the rich qualitative experiences of library users and workers.

I returned to NXL in early June 2017 to do a follow-up research interview with key participants Gill and Kathy, wanting to find out how the intensification of austerity had impacted on NXL in the intervening years, to compare this with data I had gathered in another library site in Birmingham in 2016. I found them surrounded by Labour Party ephemera and campaign materials and engaged in conversation about the next day's early-morning canvassing tasks. It was a matter of days until the 2017 snap general election and great hope and effort was being channelled into the campaign for the socialist Labour candidate of Jeremy Corbyn to win, with a view to the ending of neoliberal austerity and reinvestment in public services for the social good. Kathy and Gill reflected on how hard the last few years had been for the library, with fewer volunteers helping to keep it going and more casualties of austerity coming through the doors needing critical care and support, particularly with needs around access to digital welfare services. They narrated a shift in their activist tactics and identity, having moved from animating the library with radical cultural events and activities in previous years, to the more implicit activism of caring for people in crisis during daytime service, and channelling political energies into Labour Party organising after hours.

> We thought we'd be running a library – we never thought we'd be doing food bank vouchers. None of this. It's definitely made me more of an activist. […] Who knows, maybe we'll get a new government, this is my hope. When we see a new government, or we get a new mayor, who knows what might happen with libraries in Lewisham. I'm always hopeful that they might find some money to put a librarian in here. […] I don't know. I hope. You know, you just want things to change, don't you? (Kathy 2017)

> I'm not a do-gooder, I'm an activist. […] You can't spout about it and not do anything. I've got two little grandchildren and you just think, well what do I want out of life now? Well really, I want a better future for them. […] I have to carry on fighting. (Gill 2017)

Kathy and Gill were adamant that their commitment to keeping the library going was not an act of the 'Big Society', but part of a wider civic struggle for a fairer future. They had kept the library doors open and transformed what it could do, in order that it could one day be returned to publicly funded ownership, staffed by paid, professional librarians.

I left this interview with another formal 'exit' from this field site. Being there again and listening to its stories at this juncture of significant political change, however, reanimated my relationship to it, and although I left the site in order to focus solely on writing up the field research, NXL was still in my life on the periphery as a local resident. I could see it from my office window at Goldsmiths, and I would take the sense of it home with me each night as I processed the data and reflected on its meaning.

2018–19 Re-entry and exit

In July 2018, over a year into another term of Conservative Party government, I finally submitted my doctoral thesis. I wanted to celebrate this moment which I thought would never come and organised a garden party to celebrate. Various PhD friends came to my home for this celebration, including Kathy and Gill, whom I decided to invite in honour of the contribution they had made to my work and the knowledge I had gained from the project. Having these key participants visit my home felt both strange and familiar, as they had already been there in mind and spirit as I analysed and wrote about my ethnographic time with them.

In the autumn I began teaching at Goldsmiths, where once again I could see NXL from my office window. I would often call in there after work, donating books and catching up with Gill and Kathy. They were still struggling with exhaustion and lack of volunteers and resources and convinced me to join as a volunteer member of their management committee, attending monthly meetings to help with organisational governance and fundraising strategies. At the same time, when it was thought that local authority budgets couldn't get any worse, a further round of swingeing funding cuts came in, threating the existence of the borough's three remaining professionally staffed statutory public libraries. Another Save Lewisham Libraries campaign ensued, and I soon became its co-leader.

Once again NXL became a hub of civic activist organising, as the campaign group met there in the evenings to plan strategy, and a public campaign event was held there drawing a large, diverse crowd engaged in the continued effort to defend the borough's libraries as vital public goods. I presented some of my research findings at this event, a moment at which my dual identities as scholar and activist converged. My status as a newly inaugurated Doctor of Library Crisis added some clout to the campaign, opening doors to speak with local politicians and journalists, as well as providing a platform from which to be heard in council chambers at scrutiny committee hearings. The campaign achieved a small victory with the pausing of the library cuts for that financial year.

2020 Coda: final return and exit

In autumn 2019, tired and somewhat burned out by juggling civic activism with precarious hourly paid academic work, I left Lewisham and moved to Brighton to start a job at the University of Sussex Library. This was to all intents and purposes a final exit from the field of NXL and its environs. An opportunity to return to the field presented itself in the summer of 2020, however, when I was invited to co-write a book chapter on the impact of COVID-19 on public libraries. Not having yet gotten to know the public libraries of Brighton and Hove, it made sense to return to Lewisham to re-engage research participants there (albeit virtually), to find out how the pandemic had affected its libraries. My Zoom call to Gill and Kathy returned a gloomy picture. NXL had closed its doors to the public earlier than most public libraries, as they had no funding for security or personal protective equipment, and the health vulnerabilities of its ageing volunteers posed too great a risk. Their ability to run and resource the library had already run into decline, and the space had become mainly taken over by a credit union financial service. The concentrated crisis of coronavirus compounded the protracted crisis of austerity, incurring loss upon loss to the lifeworlds of libraries (Corble and van Melik 2021). I left this interview with a definite sense of finality this time: an exit to end my recursive series of ethnographic exits.

Discussion: oblique, open and entangled relationality to the field

My relationality to the field does not easily fit into methodological discussions of researching 'up' or 'down', 'inside' or 'outside' in the field. Rather, it is what Røyrvik (2013) calls oblique ethnographic relationality, where the researcher is positioned at the boundaries of her objects of study, traversing and studying them in a non-linear and abductive way. '[O]blique ethnography is challenging and rewarding in terms of access, entries and exits, and might be considered, in some senses, as a continuous process of entries/exits, yet offers novel research opportunities' (Røyrvik 2013: 73).

My series of entries and exits to and from NXL over the decade tell a story of an entangled and symbiotic relationship between the shifting statuses, powers and practices both of myself as a researcher, volunteer and activist and of the operations and meaning of the library itself. In the first entry and exit phase of 2013–14, my dual roles of ethnographer and library volunteer often became merged, with the demands of volunteering and participating sometimes eclipsing the attention needed for ethnographic observation. As Garthwaite (2016: 61) has surveyed, while studies of the roles of volunteers in organisations exist, there is a distinct lack of literature on the process of volunteering as a method of participant observation. Garthwaite's own ethnographic experience delivering food bank services as a volunteer addresses this gap and reveals how '[a]dopting the different roles of ethnographer and volunteer means it is inevitable that the boundaries between those roles would at times become blurred', yet, when balanced with ethical diligence and 'a stance of curiosity and openness to the unexpected', these entangled and liminal positions can produce valuable self-reflexive insights into under-researched phenomena (Garthwaite 2016: 67–68).

The interview excerpt from NXL user Jackson reveals that he referred to me as of equal standing with him and other library users and volunteers as 'ordinary citizens ... sticking together' rather than career librarians. My simultaneous subject positions of ethnographic researcher and library professional had receded into the background of how he related to me. The 'interview' conversation we were engaged in was another opportunity for 'exchang[ing] like-minded views with his fellow citizens', which for Jackson happened in the emancipatory space of the unofficial library. Reflecting now on how I related to people like Jackson, Gill and Kathy, I could be said to have over-identified with them, since social and professional hierarchies were collapsed in the radical and unstructured space of the People's Library; I was indeed 'there because I wanted to be', despite also having an agenda and obligation as a researcher. It is this volitional aspect that impelled me to return again and again. As fellow volunteer Jennifer put it, the relationality was an 'open-minded' and 'open source' one, and this open-ended process of symbiotic understanding and evolution contributed to my increasing entanglement in my attachment ties to the site. It also shaped my shifting political subjectivity and practice, as the volunteering experience helping Gill and Kathy with Labour Party organising and administration led me to join the party. Not only did 'openness' become a key emerging theme from participant observations and interview narratives, but my orientation to the field and understanding of my position in it also became opened up. My entry to the field came with certain normative assumptions.

Before my ethnographic experience, I shared the view of many library professionals and scholars that transferring the running of public services to the labour of volunteers is a 'bad' thing through its exploitation of goodwill, contribution to deprofessionalising public librarianship and complicity with neoliberal conservativism through austerity localism (Forkert 2016). My ethnographic orientation became open to a different way of understanding what was going on there and making a valuable counter argument to this, insofar as volunteering can also be a form of activism. My position as an 'ordinary citizen' participating in an open space, which was as much a site of resistance and liberation as one of neoliberal complicity, became the central way in which I related to the field and my participant-collaborators. As Marcus (2013: 206) identifies, this mode of ethnographic knowing can be thought of as lateral or even collateral, in the sense of thinking alongside or with one's participants or collaborators in the field. This requires 'a recursive rather than reflexive accounting of research as an ethnographic form' (Marcus 2013: 308, drawing on Annelise Riles' work). As my relationality to the field unfolded over time, recursivity became key to the abductive ways in which I made sense of the field and my position in it, both within and beyond the PhD.

When I returned to the field in 2016–17 after a PhD hiatus – first in a non-researcher capacity, as a temporary local authority library employee – the boundaries of what constituted my field and place in and understanding of it became 'fuzzier' (Nadai and Maeder 2005). While this period of employment did not constitute fieldwork in an ethnographic or formal research sense, it nonetheless informed my understanding of the field and added further layers of meaning to my identity as a library practitioner and thinker. It also further complicated what it meant to be inside or outside the field, not only due to my shifts in status passage, but also due to the unstable organisational ground of the (sectoral) field itself. In this way, my experience in 'the field' has been methodologically messy, confusing my positionality and analytical frames. However, such mess is arguably both productive and performative of the phenomena under analysis (Law 2004). Marcus' methodological notion of circumstantial activism is helpful here, which he defines as ethnographic activity that moves in recursive and participatory circuits, caught in normative binds of 'loyalty, sympathy, and affiliation', which must – rather than simply declaring the researcher normativity involved in this – instead 'make trials or contests of norms and values the basis of forming working collaborations and arguments, with uncertain, often messy outcomes in pursuit of ethnographic insights in the field' (Marcus 2013).

The ethnographic insights I gained from returning to the field and re-interviewing Kathy and Gill in 2017 as they occupied the knife edge of fighting for the library's survival through campaigning for political change included a key observation that while I had set out to study a library, what became the object of study was how this library was manifested in the actions and interactions of two women and their community. I realised that libraries are made of people above all else, and as one of those people who constituted the NXL community I shared an ethic of care for its welfare and future. In the following year, my relationality to these key participants became even more blurred as they stepped over the threshold into my home (both intellectually and literally) and I in turn stepped over the threshold from researcher to library volunteer manager and campaigner. As Gill put it in the interview extract above, I couldn't just 'spout about it', I had to do something.

My compound identity as a scholar, activist and library professional operated as important axes on which to articulate and defend library value. Reflecting on this now, the lessons I had learned about how to do community and political organising had come through observing and working with seasoned citizen activists Kathy and Gill and other members of the NXL community, an educational experience that was of much greater value and impact than the scholarly lessons of the academy; yet it was the power and authority of the PhD letters after my name that helped my voice to be heard in the institutional spaces of the council and the press. This reveals a shift and tension in the power dynamics between me and my research participants and library collaborators. When I was immersed in the library as a volunteer in 2013–14, despite having a research agenda, this had faded into the background and I operated on a more equal footing with fellow library volunteers and users. When my professional status as a doctor had become more foregrounded post-PhD, this allowed a certain kind of platform denied to many fighting for library and personal survival. I aimed to balance this by attending in self-reflexive and collaborative ways to how I could mobilise the knowledge I had gained through the research journey for the benefit of those who helped to inform and shape that knowledge. There is a difference, then, between using one's power as a means to an end and sharing forms of collective empowerment through co-produced knowledge.

My brief and final return to and exit from NXL in 2020 revealed that the radical project of the People's Library had ended, with the pandemic driving the final nail in the coffin, on top of four waves of austerity cuts. I left the field this time not out of choice, but because it was no longer there. However, its impact on my identity as a library practitioner, researcher and activist is very much still alive. I have carried the lessons of the previous decade of entanglement in this site with me into my new place of employment and civic life. I was able to apply the knowledge and skills of community and political organising and engagement gained through NXL to trade union activism in my new employment, as well as co-organising mutual aid during the COVID-19 crisis in my new local neighbourhood.

Conclusions: three methodological lessons

My experience of entering, leaving, returning and orbiting the field in a recursive cycle of encounters and status passages was generative for producing colateral methodological and activist praxis, but it also brought with it significant challenges in the writing-up process, since the boundaries of the field were difficult to discern and draw. Each time I returned to the field it had shape-shifted again, and so had I. Despite these challenges, however, my multi-sited ethnography proved a worthwhile endeavour and a fitting way to address my overarching research question about the double bind of social and organisational change that public libraries are caught up in.

Three concluding lessons can be drawn from my entanglement in and extraction from this fuzzy field of research.

Recursivity increases fuzziness but enhances reflexive insights and impact

Repeated and entangled returns to the field pose risks as well as benefits for research processes and outcomes. I was potentially too open or too close to the field and its subjects; however, this openness was both performative and productive of the objects of analysis. Entanglement and seemingly eternal returns can lead to information overload, exhaustion and burnout, as well as boundaries that are too blurred and fields that are too fuzzy. This can lead to an inability to know what to follow or focus on within the field, and also runs the risk of blurring ethical boundaries by caring too much or bringing too much (inter)subjective normativity into the research. My resultant doctoral thesis became rather unwieldy and incorporated a huge range of concepts and theories to make sense of the data and experience I had amassed. A lesson here is to limit the number of theoretical frames for abductive fieldnote analysis, in order to focus more on depth than breadth in how to write about such a protracted research journey. This can enable a way to focus in on key areas of the fuzzy field, sharpening reflexive insights and enabling more impact for the social problems the work seeks to address. Taking breaks, coming and going to and from the field enables the researcher to notice more changes over time. If I'd been there constantly or in one demarcated period of time only, I would not have noticed or understood social change in the same way. It enabled a different kind of witnessing, which included witnessing significant loss (Robinson and Sheldon 2019) as well as seeing participants age and gain hope or wisdom.

Embrace and explore contradictions

After 'exiting' the field in a formal ethnographic researcher sense in 2018–19, my dual roles as NXL management committee member and Save Lewisham Libraries campaign co-leader were somewhat in tension, since on the one hand I was fighting to save professionally staffed public services, while on the other I was complicit in upholding the alternative volunteer model that supports austerity localist models of outsourcing to precarious communities (Forkert 2016). However, my research found that volunteering and activism are not mutually exclusive, particularly in cases where acts of resistance to neoliberal hegemony can occur through both the implicit and everyday acts of caring for fellow citizens in crisis in 'meantime' third spaces of service provision (Williams et al. 2016), and the more explicit activist acts of inviting Black Panthers and socialist film-makers to engage with library publics, or campaigning for local and national political change through the library's networks. I also came to learn through my decade of entanglement with NXL that embracing or inhabiting contradictions can be an inherent part of praxis-based scholarly activist commitments. But this insight is more easily realisable once a clear ethnographic extraction from the field has been achieved. Volunteer-activist research is a learning process. In this recursive and contradictory journey, I learned that we should challenge and change the ways in which both libraries and volunteering as activism can be conceived and practised.

Moving between closed and open: towards praxis-oriented research

Open and oblique relationality and embracing contradictions in the field through recursive entries and exits teaches us about shifting power dynamics vis-à-vis the nature of scholarly activist praxis. As the discussion above demonstrates, the power relations between myself and my participants shifted over time as I became empowered both through what I learned from my activist mentors at NXL and through the doors opened by achieving the PhD. Coleman's work on ethnographic exit points as epistemological hinges via which we can occupy the gaps between theory and activist practice is helpful here. She considers how exiting and analysing the empirical field, rather than closing off activistic practice, opens up a field of knowledge production as one moves in an oscillating fashion between theory and practice, in a

> persistent back-and-forth movement between committed engagement and an ethos of critique. […] While commitment may entail 'closing the gaps,' critical engagement requires us to persistently reopen the breach: to detach from the processes within which we are engaged to see how resisting subjects are made, how struggles are managed, contained or enmeshed within relations of domination, and how our own knowledge practices are likewise a product of our imbrication within those power relations. (Coleman 2015: 277)

The social justice actions I went on to undertake in the field beyond my research commitment were a product of praxis-based learning and collaboration. This work naturally had its limitations, but the lessons I learned in the field have a legacy in my ongoing transformation as a library researcher, practitioner and activist.

These three lessons are valuable for other ethnographic projects and future possibilities for collective action-based research. Perhaps the writing of this chapter is a final exit from this particular field, leaving me furnished with greater methodological insights for my next circumstantial activist endeavours.

References

Coleman, L.M. (2015). 'Ethnography, commitment, and critique: Departing from activist scholarship'. *International Political Sociology*. 9(3): 263–80.

Conrad, M. (2010). 'Lewisham asks community groups to run libraries'. LocalGov (blog). 6 December 2010. www.localgov.co.uk/Lewisham-asks-community-groups-to-run-libraries/33791 (accessed 18 January 2023).

Corble, A. and van Melik, R. (2021). 'Public libraries in crises: Between spaces of care and information infrastructures', in R. van Melik, P. Filion and B. Doucet (eds) *Public Space and Mobility*. Vol. 3. Bristol: Bristol University Press.

Forkert, K. (2016). 'Austere creativity and volunteer-run public services: The case of Lewisham's libraries'. *New Formations*. 87(87): 11–28.

Garthwaite, K. (2016). 'The perfect fit? Being both volunteer and ethnographer in a UK foodbank'. *Journal of Organizational Ethnography*. 5(1): 60–71.

Law, J. (2004). *After Method: Mess in Social Science Research*. London: Routledge.

Marcus, G.E. (2013). 'Experimental forms for the expression of norms in the ethnography of the contemporary'. *HAU: Journal of Ethnographic Theory*. 3(2): 197–217. https://doi.org/10.14318/hau3.2.011.

Nadai, E. and Maeder, C. (2005). 'Fuzzy fields. Multi-sited ethnography in sociological research'. *Forum Qualitative Sozialforschung/Forum: Qualitative Social Research*. 6(3). www.qualitative-research.net/index.php/fqs/article/view/22.

Robinson, K. and Sheldon, R. (2019). 'Witnessing loss in the everyday: Community buildings in austerity Britain'. *The Sociological Review*. 67(1): 111–125.

Røyrvik, E.A. (2013). 'Oblique ethnography: Engaging collaborative complicity among globalised corporate managers', in C. Garsten and A. Nyqvist (eds) *Organisational Anthropology: Doing Ethnography in and among Complex Organisations*. pp. 72–88. London: Pluto Press.

Williams, A., Cloke, P. May, J. and Goodwin, M. (2016). 'Contested space: The contradictory political dynamics of food banking in the UK'. *Environment and Planning A: Economy and Space*. 48(11): 2291–2316.

12

On the importance of intermissions in ethnographic fieldwork: lessons from leaving New York

Joe Williams

Introduction

In the pages that follow I discuss the notion of 'intermissions' in ethnographic fieldwork. By an intermission I mean a temporary physical or geographical exit from the field, with the intention of returning; a breaking from the place in which the work of ethnographic observation happens. This intermission, for me, came at the halfway point of my fieldwork; it was a leaving of the field and the country in which observation had taken place, and was accompanied by the presentation of 'initial findings' at a conference. The presentation was followed by a discussion among the audience about homelessness, what it was, who it involved and how it should be studied, measured, attended to and talked about. This discussion involved a formulation of homelessness as something conceptually clear and identifiable to sociologists outside of the field, outside of the context of study. I felt that my own presentation facilitated this decontextualised talk. The talk did not match my actual fieldwork experience; it did not match the actual occasions in which the outreach workers I was supposed to be observing dealt with the notion of 'homelessness'.

The chapter is organised around two sets of description, of distinct form: the first is of actual fieldwork experience (an ethnography of homeless outreach teams in Manhattan); the second is a description of some methodological considerations in dialogue with an ethnomethodological sensibility. The intention is to match my own fieldwork experiences, which included an intermission and a rethinking of a PhD project, with a methodological reflection on this process. The focus of the chapter is on a moment in which some analytic troubles were made clear: namely, that the social practice I was trying to deal with *as* a social practice – homeless outreach – is clouded by the ways in which 'homelessness' is commonly discussed and is taken for granted. The consequence is that homeless outreach is assumed to be dealing with a clearly predefined and readily visible object in the shape of 'the homeless'. The analytic trouble, following the intermission, is resolved by representing such practices as 'occasioned' rather than as theoretical constructs, which avoids reproducing decontextualised and assumed understandings of topics as social problems rather than as accomplished through actual practices. In this sense, the chapter is a methodological contribution that places an emphasis upon a somewhat neglected reflexive process that can accompany an intermission in fieldwork, and the possibility that the process poses for a considerable reworking of a project and the way in which a social problem might be formulated. While I borrow much from ethnomethodology here, it is not my priority to champion this particular programme. Instead, the intention is to attach my fieldwork experiences to some already laid-out notions of 'respecification'.

'Visibility' is something of a key mediation for this project. With homeless outreach, the ways in which visibility gets done as an activity allow categories of need to be seen and attended to. Visibility enables both the generating of the category and its subsequent detection, for practical purposes at least. Detailed discussions and accompanying ethnographic fieldwork exploring this relation in outreach work exist elsewhere (see Hall 2017; Hall 2018; Smith 2011), and so the unfolding of the concept will not be done in this chapter, although it is broadly relevant. Rather, my focus remains on the fieldwork intermission that occasioned the conditions in which the *who* and *how* of visibility of the topic of study became significant. Significant, that is, for the outcome of a research project and for the reflection on the role of the researcher within the field. Perhaps even, at a stretch, for the possibility of the 'field' itself.

I start by expanding on these 'actual occasions' of fieldwork and outreach work.

The rhythms of fieldwork

I begin with the fieldwork, an ethnography of homeless outreach teams in New York, in downtown and mid-Manhattan, mostly on sidewalks, handing out food to people from the back of a van. I was to be there for one year, a duration that was determined by funding conditions, a visa expiration date and an early (perhaps overly literal) reading of Erving Goffman's 'On Fieldwork' (1989: 130) in which he suggests spending 'at least a year in the field. Otherwise ... you don't get deep familiarity.' The 'field' was anthropologically clear (geographically and temporally) but sociologically blurry (exploring 'homelessness' as a subject). The exploration of this sociological blurriness was, of course, a central part of the fieldwork process. My immersion into the practice of homeless outreach began the day after I arrived in Manhattan, and went on to take various forms which lent themselves to different kinds of exposure to homeless outreach and what could be (and certainly is) described as 'homelessness'. I shall give some detail.

Homeless outreach, on the street, is not a standardised practice. Yet, comparing my observations with those of others (Hall and Smith 2017, for example), practitioners tend to share some rules. You do not wake people who are sleeping. You do not judge people. You do not force assistance on those who don't want it (although there are degrees of interventionist approaches that mostly depend on what kind of authority one might wield). Lastly, you do not give too little and you do not give too much. Too little is not enough, too much may go to waste and is more efficiently used elsewhere. The list of 'do nots' was largely shared among different outreach groups. The 'dos', however, the methods for delivery of assistance, are where things start to become more diverse.

To explore the diversity of methods involved in 'doing being an outreach worker', I arranged to meet and participate with several outreach groups. A group could mean anything from a team of twenty people to only myself and one other. The level of organisation involved in an evening's outreach would also change. Some groups would have a carefully arranged and prepared agenda, keeping to a strict schedule, whereas others would adopt a more sporadic movement through the city. One of the first groups I joined was run from a small shop front upstate.

For me, the evening would involve travelling upstate to help with preparing food and supplies, loading a van and a short briefing, before driving down into Manhattan and following a series of prearranged stops in the downtown area. Outreach workers and volunteers would supply homeless clients with a range of clothing items, hot drinks and food, sandwiches for the following day and sleeping bags and blankets for outdoor/subway sleeping. The applicability of the category 'homeless' was confirmed by the question, 'Are you homeless?' The right answer to get food and supplies was 'Yes'. When the planned route was completed we would drive back upstate and unpack the van, and I would catch the train back to the city in the early hours. The outreach group consisted mostly of volunteers, some of whom participated regularly and others less so. As such, some volunteers became familiar with regular users of the services, whereas others were essentially strangers.

Another group I joined included regular teams who would volunteer multiple times a week and included 'workers': professional outreach staff. Such teams would do much the same stuff as the first group described, but they came to know their 'clients' well. They were on first-name terms, would know what their clients' preferences were from among the food and supplies they had to offer, and became well acquainted with the different and particular personal difficulties that clients might experience. They would arrange their practice to cater to certain needs and would save food for particular people who they knew would most appreciate it. Further, they would arrange their outreach practice according to what they came to understand as a kind of hierarchy, and an urgency, of need. Again, the schedule they kept was quite precise, visits to locations were bound to certain times and the 'consistency' of their service was considered an important part of their work, providing some small measure of certainty to their clients. No matter the day, weather, traffic, they would be at their 'stops' throughout the city.

With both of the above groups, homeless people, clients, strangers who waited for the service at the stops, or who walked up unexpectedly, would be confirmed as 'homeless'. Visibility of need plays a significant part here. However, that does not necessarily mean that those people were homeless in any literal sense. The term 'homeless' as used by the outreach teams refers to a range of circumstances (observable in a number of ways) and is often considered an adequate description (rather than literal) and a justification for the kind of care work that these groups do. A client may very well come to the attention of the outreach team by being literally homeless; however, that situation often evolves, with clients becoming housed, entering a temporary shelter, finding some place to be that is not necessarily the street. In most cases, this does not mean that there is a reduced urgency of requirement for what outreach teams offer. Being 'homeless' moves through meanings. This movement became a priority of my observations.

Fieldwork continued in directions which fell outside of what I suspected outreach would look like. A 'group' one night was just one other person. He worked in the city. Having finished at the office, he swung by a store to buy blankets, sandwiches and bottled drinks, with which he filled his car. We had arranged to meet at a downtown intersection and spent the first part of the evening driving (and driving fast, in his convertible BMW) between some 'hotspots' for homeless people. The evening turned into night, and we still had plenty of supplies left. It was 11:30pm when he said, 'I'm just going to drive around until I've given everything away. You in?' 'Sure.' We drove until 3am, both of us scanning the sidewalks, doorways and benches. Whenever he saw someone who 'looked' homeless, he would shout, 'There's one! Joe, get them a blanket!' I would jump out of the car and hand the person a blanket and some food or bring them over to the car and they could choose what they wanted. It was a frenzied effort, and one that the recipients were not expecting, but mostly not averse to. On this one evening there were several occasions when we interrupted people on the street who were mid-drug use, engaging in sex acts or washing themselves. It would be unlikely that the more organised outreach efforts would encounter their homeless clients in the same way. The point here is that outreach work, and its social resources, included a varied range of practices and people. It was my task as an ethnographer to observe how these practices were assembled.

The methods for doing outreach work differed, and so too did the clients. Outreach workers make attempts to organise clients based on familiarity with that person and their habits, their particular kind of need, the urgency of it and where they are encountered. However, the why and how of a person's arrival at being homeless in New York can be not only an individual set of circumstances but 'occasioned' by the encounter, the interaction. The problems commonly associated with homelessness, such as addiction, mental health issues, financial problems, unemployment, lack of networks of support, are not solid explanations for people's becoming homelessness. They are not useless or unused explanations, but their use, for the practical accomplishment of outreach work, is always of the occasion, of that particular moment. This is not only because often people are in the midst of a combination of issues, and the prevalence of one issue over another might be seen at different times, but because from one night to the next clients, homeless individuals, observably and through interaction, move in and out of particular categories of need. Further, some clients present different reasoning on a daily basis for their homelessness or for anything else. Others are unreliable in different ways. As such, what it means to be 'homeless' becomes an elastic notion, and outreach workers organise their practice to attend to this elasticity.

This brief description of my fieldwork in the first six months, the first half of my PhD project, highlights that while on the one hand the project and the 'field' were clear and easy enough to describe and talk about, representing homeless outreach and defining homelessness was not a straightforward sociological task. I had reached a point in which I was looking for a way to deal with my observations and fieldnotes so that I could attend to the 'elasticity' of homelessness and of homeless outreach practices. I wanted to show that homelessness does not mean something intrinsic about the situations in which it is established, whether that be outreach workers' encounters with clients or academic publications (I am referring here to the kind of textual representations discussed in Atkinson's (1990) *The Ethnographic Imagination*). The understanding of the category doesn't carry from one situation to the next, not perfectly intact. It is re-established each time, indexed by familiarities, similarities and differences.

The intermission

Cue the 'intermission'. I left New York and returned to Cardiff and my home university, the purpose being to meet with my supervisors and review the first half of the fieldwork. This hiatus was a condition of my funding, to check my progress, discuss any emerging lines of analysis and prepare for the final stage of the project. The return to Cardiff also coincided with a postgraduate research conference at which I planned to give a presentation on my 'initial findings'. At this point of the fieldwork, coherently describing the difficulty of the elasticity of homelessness was not something I was prepared or in fact able to do. Instead, my presentation revolved around what homelessness in New York looked like, using various forms of data, some fieldnotes, tables and graphs, and quotes from other current research. I was unhappy with my own presentation of the topic, as it lacked any means of accessing the complexity, the occasioned-ness, of the category 'homeless' that I had observed. I presented homelessness as a social problem, something generalised, broadly defined according to my own, not my informants', parameters. The short discussion that followed the presentation was similar in tone. Members of the audience used their own understandings of homelessness as relevant to the discussion of homelessness as a social problem, with possible causes and solutions and guaranteed features. The representation was a theoretical one or, rather, one that was bound up in a felt need *to* theorise, instead of in keeping, with the actual observations that composed the fieldwork.

Following the conference, I expressed my dissatisfaction with how my approach to presenting the topic had proved to distort the phenomenon, my frustration that my theoretical and decontextualised understandings (obtained from reading on the 'topic' of homelessness and urban ethnography) *of the field* were not matching what I was observing *in the field*. I began to explore different ways of considering the understanding and representation of an observed reality. This included an over-confident first reading of Wittgensteinian philosophy that quickly left me feeling unmoored. It did, however, lead to a more careful look at some phenomenological readings, and eventually I focused my attention on some of the sensibilities of ethnomethodology. There is a logic to this choice of focus, and an accompanying caveat to some of what I have already said. I was interested in the *practice* of outreach work, the way in which encounters between service providers and clients happened. This oriented my sociological handling of observations to interactionism (I leave this broadly defined for now). Ethnomethodology directs much attention to the study of, and ontological dealing with, practical accomplishments. As such, my comments regarding members of the audience of my presentation misrepresenting the phenomenon are not to say that they were 'wrong' to do so – much of what was said could be insightful for a study of another kind; my focus on practices made it clear that my study *had* to deal with specific occasions, rather than theorised concepts.

The intermission, geographical and reflexive in nature, provided the circumstances in which I began a reworking of my approach to the field.

Being required to present my research before it was complete, and taking time to reflect on my own formulating of the subject of study, created a space for me to consider how a distortion of the topic might be happening. Further, it provided an opportunity to adjust my approach before re-entering the field. Here, then, is the essential 'lesson' to take from this chapter regarding leaving the field: that the factoring in of an intermission in ethnographic fieldwork might allow for an essential moment of reflexive thinking and reformulating of an approach to study. Moreover, and more specifically, the requirement of presenting data can be a useful method of accounting (to yourself and others) for the way in which the topic is being considered.

The suggested principle of an intermission is a simple one. What follow now are some of the particular details of my own reformulation of the topic of study and of my overall approach to the 'field'. As noted previously, this is largely borrowing from ethnomethodology – its priorities and sensitivities – and can be read as a very brief overview of the ethnomethodological take on the 'status' of the observer. I have attempted to be concise and economical with both words and principles.

Return to the field

The return to the field was accompanied by an intentional shift from a topicalising of the field via sociological theorising to observing how members of society are topicalising the field through their own localised practices. I became concerned with showing how members develop their own logic by which their practices, order and correctness can be judged *in situ*. The way of doing this was to take a close look at how people are making sense, and making sense visibly, in and through the actual occasions observed, and how they are collaborating with each other to produce an account of the world (Garfinkel and Sacks 1970).

I identified that the notions of homelessness in my initial observations consisted of my own topicalising of the subject, so I sought to reformulate the parameters of the phenomenon, locating within my fieldnotes and observations where homelessness was being topicalised by my informants for each other as members. In this way, the methods for discovering the phenomenon became the same as the methods for producing the phenomenon; being a member of that society and competently using the social resources which are available. This represented another shift, one which sees members of society not necessarily as experts in the culture of which they are a part, but as fellow enquirers into it (Sharrock and Anderson 1982). It is then the everyday methods that people use in accomplishing mutual understanding that enable the 'culture' to be a reportable and discoverable phenomenon.

The ethnographer gains access to these understandings through developing a similar competency in those methods. The analysis turns to showing just *how* it is possible for those interactions to be recognisable and intelligible to members; through their (and your) competent use of natural language, and natural language activities (Garfinkel and Sacks 1970).

There is then, a sort of double sociology to do; first, the 'lay' sociology that members of the situation are doing (outreach workers and their clients ordering actions and talk into observable scenes). The second kind of sociology seeks to explicate the ways in which these practices are assembled, and is done through sociological description (fieldnotes and ethnographic writings being one way). This could be simply described as the process of making observable the ways in which the members make things observable. This is referred to as 'ethnomethodological reflexivity' – the reflexivity operating at the root of every situation. Garfinkel and Sacks (1970: 338) emphasise the following:

> Reflexivity is encountered by sociologists in the actual occasions of their inquiries as indexical properties of natural language. These properties are sometimes characterised by summarily observing that a description, for example, in endless ways and unavoidably, elaborates those circumstances it describes and is elaborated by them.

In discussing these kinds of analytical techniques, Sean Rintel (2015: 125) frames it this way: 'Categories such as gender or relationship are often treated by researchers as inescapably there *to be found* and amplified in importance rather than found to be relevantly occasioned.' In my own observations I treated 'homelessness' as an already defined concept and as 'there to be found'. I realised that I had oriented my ongoing analysis of collected materials to amplifying its importance as I *already* understood it. This proved to eclipse a noticing of members' own formulation of 'homelessness' as a resource for intelligibility within interactions. Garfinkel (1991) addresses this, using the term 'haecceity' to refer to the properties or quality of a thing which is unique or 'just this one' in 'just this situation', avoiding suggesting that topics have a stable 'core'.

What was important for my return to the field was that members are reflexively orienting to the observable order of everyday actions and talk, and this enables the ethnographer to go beyond just an 'ethnographic reflexivity', going further than considering how the researcher's methods are distorting the phenomenon or the setting, and to see how members' methods are already constructing and negotiating the setting. The fact that members are already making observations, for each other to discover, about the topics which are relevant provides an intentionally discoverable phenomenon by which the setting is already being described, topicalised and reflexively referred to. Essentially, the members' own methods produce the field.

The ethnographer as another member

Given my study of outreach practice, it was the focus on the *practical* implications of indexical expressions that was of interest to me. The practical, rather than theoretical, interest that *members* have in their constitutive work. The indexicality of expressions refers to talk that is understood in reference to just what is going on in that particular occasion. For my own observations going forward, I tried to notice what was understood by outreach workers when they talked about 'homelessness', and what it meant *for* their way of doing outreach work, while at the same time *how* their particular way of doing outreach work formulated meanings of 'homelessness'. The description elaborates the circumstances, but the description is also elaborated by the circumstances. Further, the description does not stand outside of the circumstances it describes, it is oriented to by members so as to organise and make sense of the situation. How, then, to consider myself, a 'participant-observer', within these situations?

To a sociologist looking to read into ethnomethodological understandings, 'respecification' is a useful term to keep close (see Garfinkel 2002). The concepts found in Western science and philosophy, such as order, logic, rationality, action, are respecified as *members'* practices (Lynch 1993). For sociology, a particular respecification of interest is one of Durkheim's aphorisms (originally found in *The Rules of Sociological Method* (1938)):

> The objective reality of social facts is sociology's fundamental principle.

Garfinkel and Sacks (1970) offer this adjustment:

> The objective reality of social facts is sociology's fundamental phenomenon.

That is, the concern is for not substituting 'objective' expressions for the indexical properties of members' practical discourse, but to examine the rational accountability of everyday actions. The 'knowing' by members is achieved through the exhibiting of the account (talk and action), and the account organises the understanding of the setting (reflexively orienting to that account), and this is considered as the 'objective reality of social facts' (the occasions with which the field is made). So, it stands that sociological discoveries are, in every case, discoveries *from within* society (Turner 1971: 177 cited in ten Have 2001) inasmuch as the researcher, the same as any observer, uses their membership knowledge to understand the materials produced. As Paul ten Have (2001: 36) summarises:

> The notion of member refers to capacities or competencies that people have as members of society; capacities to speak, to know, to understand, to act in ways that are sensible in that society and in the situations in which they find themselves.

The implication of this 'notion of member' for ethnography is that, in order to be able to study the specific details of a particular practice, one must develop a competence in doing that practice oneself. Garfinkel refers to this as 'the unique adequacy requirement' (Garfinkel and Wieder 1992:182; and see Smith (in press)). This is the requirement for the analyst to be at least 'vulgarly competent in the local production and reflexive natural accountability of the phenomenon of order [he] is studying'. In essence, to be another member of that society.

This notion of member might be compared to another, familiar concept, that of the 'Native'. 'Member' and 'Native' has been considered in a recognisably similar way to what is being outlined here. Blumer (1969: 542) suggests that 'one would have to take the role of the actor and see his world from his standpoint'. This is not a far cry from Garfinkel's 'unique adequacy requirement'. However, Alex Dennis (2011: 349) offers this comparison;

> For symbolic interactionists, the sense of interaction depends entirely on actors' interpretations and understandings (Blumer 1969: 2), while for ethnomethodologists the meaning of any interactional 'move' is reflexively tied to its context: action, sense, and situation are mutually elaborative in situ (Garfinkel 1967: 3–4).

Both approaches can be considered as an analytical apparatus and as a means of developing sociological descriptions. The crucial point here is a consideration of the 'native' not necessarily as an expert in the culture of which they are a part, but a as fellow enquirer into it (Sharrock and Anderson 1982). It is the everyday methods that people use in accomplishing mutual understanding that enable 'culture' to be a reportable phenomenon. The ethnographer gains access to these understandings through a competency in those methods.

A useful comparison might be found in considering the 'common danger of ... going native' (see Hammersley and Atkinson 2007: 87). The 'danger' here is in reference to the potential for the abandonment of the analytical task in favour of 'the joys of participation', or a bias forming and leading to 'over rapport'. The essential concern raised here is that a *problematising* (by the analyst) of members' perspectives will be missed. Consider this an important point of distinction, that the analytical task with which members are concerned is done in the first case by members, and second (if at all) by the analyst. Rod Watson (2015: 31), writing about the work of Harvey Sacks, describes this as a 'primordial phenomenon'; it is the practical relevance of locally situated practices of social organisation and sense-making of members that is the topic of study. In this sense, the 'danger' lies not in going native but in considering the native as different from 'us' (us and them/native and ethnographer). If the native is considered in this way, it risks turning 'their' culture into a monolith constructed out of cultural differences, the typical features of which it becomes the ethnographer's task to describe, and not to partake in as a member (Sharrock and Anderson 1982). Such typical features may be adequate for comparing 'different' cultures, but not for analysing the situated practical accomplishment of the everyday life of members. For the researcher, 'going native' – especially considering Garfinkel's 'unique adequacy requirement' – is a necessity rather than a danger. What could be considered 'native', a member of a society, is an ongoing negotiation, the practices of which are reflexively accountable in the occasion rather than existing in an absolute, intrinsic state.

To conclude, in returning to field after an intermission, I sought to direct my focus to the practices, talk and practical actions of the outreach workers I was participating with. Rather than fretting about the elasticity of the object of their work, the blurry notion of homelessness, I was able to describe in exact, situated, detail how it was that workers formulated notions of homelessness through, and in order to accomplish, their practical tasks. The noticings of my ethnography were redirected away from ideas about society and sociability that were predetermined as 'there to be found' (a theoretical understanding of homelessness), and towards the occasioned reality of those whom I was there to observe and participate with.

References

Atkinson, P. (1990). *The Ethnographic Imagination: Textual Constructions of Reality*. London and New York: Routledge.
Blumer, H. (1969). *Symbolic Interactionism: Perspective and Method*. Englewood Cliffs, NJ: Prentice-Hall.
Dennis, A. (2011). 'Symbolic interactionism and ethnomethodology'. *Symbolic Interaction*. 34(3): 349–356.
Durkheim, E. (1938). *The Rules of Sociological Method*. (8th edn). Chicago, IL: University of Chicago Press.
Garfinkel, H. (1967). *Studies in Ethnomethodology*. Englewood Cliffs, NJ: Prentice-Hall.
Garfinkel, H. (1991). 'Respecification: Evidence for locally produced, naturally accountable phenomena of order, logic, reason, meaning, method, etc. in and as of the essential haecceity of immortal ordinary society (I): an announcement of studies', in G. Button (ed.) *Ethnomethodology and the Human Sciences*. pp. 10–19. Cambridge, UK: Cambridge University Press.

Garfinkel, H. (2002). *Ethnomethodology's Program: Working Out Durkheim's Aphorism*. Lanham, MD: Rowman & Littlefield.
Garfinkel, H. and Sacks, H. (1970). 'On formal structures of practical action', in J.C. Mckinney and E.A. Tiryakian (eds) *Theoretical Sociology: Perspectives and Developments*. pp. 338–266. New York: Appleton-Century-Crofts.
Garfinkel, H. and Wieder, D.L. (1992). 'Two incommensurable, asymmetrically alternate technologies of social analysis', in G. Watson and R.M. Seiler (eds) *Text in Context: Studies in Ethnomethodology*. pp. 175–206. Newbury Park: Sage.
Goffman, E. (1989). 'On fieldwork'. *Journal of Contemporary Ethnography*. 18: 123 doi: 10.1177/089124189018002001.
Hall, T. (2017). *Footwork: Urban Outreach and Hidden Lives*. London: Pluto Press.
Hall, T. (2018). 'Homelessness and the city', in S. Low (ed.) *The Routledge Handbook of Anthropology and the City*. pp. 55–68. New York: Routledge.
Hall, T. and Smith, R.J. (2017). 'Seeing the need: Urban outreach as sensory walking', in C. Bates and A. Rhys-Taylor (eds) *Walking Through Social Research*. pp. 39–53. Abingdon: Routledge.
Hammersley, M. and Atkinson, P. (2007). *Ethnography: Principles in Practice*. (3rd edn). London and New York: Routledge.
Lynch, M. (1993). *Scientific Practice and Ordinary Action: Ethnomethodology and Social Studies of Science*. New York: Cambridge University Press.
Rintel, S. (2015). 'Omnirelevance in technologized interaction: Couples coping with video calling distortions', in R. Fitzgerald and W. Housley (eds) *Advances in Membership Categorisation Analysis*. pp. 123–150. London, UK: Sage.
Sharrock, W.W. and Anderson, R.J. (1982). 'On the demise of the native: Some observations on and a proposal for ethnography'. *Human Studies*. 5(2): 119–135.
Smith, R.J. (2011). 'Goffman's interaction order at the margins: Stigma, role, and normalization in the outreach encounter'. *Symbolic Interaction*. 34: 357–376. https://doi.org/10.1525/si.2011.34.3.357.
Smith, R.J. (2022). 'Fieldwork, participation, and unique-adequacy-in-action'. *Qualitative Research*. https://doi.org/10.1177/14687941221132955.
ten Have, Paul. (2001). 'Revealing orders: Ideas and evidence in the writing of ethnographic reports'. Paper read at the IIEMCA conference on 'Orders of Ordinary Action', 9–11 July 2001, Manchester, UK. www.paultenhave.nl/RO11.htm.
Turner, R. (1971). 'Words, utterances, activities', in J. D. Douglas (ed.) *Understanding Everyday Life: Towards a Reconstruction of Sociological Knowledge*. pp. 169–187. London: Routledge and Kegan Paul.
Watson, R. (2015). 'De-reifying categories', in R. Fitzgerald and W. Housley (eds) *Advances in Membership Categorisation Analysis*. pp. 23–50. London: Sage.

13

Can you remember? Leaving and returning to the field in longitudinal research with people living with dementia

Andrew Clark and Sarah Campbell

Introduction

This chapter draws on our experiences of leaving and returning to the field in an investigation of the neighbourhood experiences of people living with dementia. Between 2014 and 2019 we engaged in a five-year longitudinal study of the neighbourhood experiences of people living with dementia and their families, friends and care partners (Ward et al. 2018). We deployed a range of approaches and methods that are rooted in ethnographic practice, placing fieldwork and the sustained, repeated engagement with participants in particular places over a period time at the centre of our approach. The 'field' we were concerned with was not simply a geographically bounded location (such as the neighbourhoods where participants lived), but also temporal (incorporating change over time) and social (incorporating relational ties with other people regardless of their location). The 'field' thus emerged from ongoing relations between us and participants nurtured over time (Caretta and Cheptum 2017). It was also constructed through intellectual inquiry and engagement with research material during periods of data collection, analysis and dissemination (Amit 2000).

Repeated interactions with participants, and their associated networks of friends, family members and acquaintances, in the places they visited or where they lived prompted a messy process of entering, 'leaving' and re-engaging with what we came to recognise as the field. The presence of dementia also heightens some of the challenges for understanding what it means to leave and return to the field at regular intervals. Our intention in this chapter is not to provide a confessional tale or how-to guide for engaging and disengaging with the field in a dementia context. Rather, by focusing on fieldwork designed to better understand the lives of individuals living with dementia, we intend to question and recast what it means to leave and return to the field.

Dementia is associated with a range of symptoms including cognitive change such as memory loss, declining physical abilities and communication difficulties. Over time it could become difficult for participants living with dementia to cognitively and physically access, recognise or locate themselves in the fields we were researching in the ways that they might have done earlier in the research. Some might also be unable to remember previous interactions with the research team or the experiences they previously have narrated to us. So, it is tempting to reduce what was happening here to a somewhat pithy perspective that it was the participants, rather than the researchers, who were 'leaving the field'. In practice this was considerably more nuanced and requires careful reflection, and in this chapter we reflect on what it means to leave, return and remember the field as a cognitive as well as a physical and temporal location. We consider how researchers might plan for and respond to the possibility that participants will experience cognitive decline or memory loss over the course of research, the ethical implications that might arise from this and the impact on what we might come to recognise as 'the field'.

Introducing the wider research context

Dementia is a progressive condition that has become a global health priority. The worldwide prevalence of dementia is set to increase from 54 million to around 130 million by 2050 (Alzheimer's Society 2021). It is estimated that 850,000 people in the UK live with a diagnosis of dementia (Alzheimer's Society 2021), many of whom will continue to live in their own homes with the assistance of family, friends and support services.

It was within this context that we sought to understand how people living with dementia, and their care partners, understood local places with which they associate. We were interested in how they made sense of the places where they lived and experienced the intersection of people, activities, history and biography when going about 'neighbourhood life'. The work was completed in three locations in England, Scotland and Sweden, with ethical approval obtained for the research across all three settings including the relevant National Health Service (NHS) Health and Social Care panel in England. This chapter is based on activities in England.

We encouraged participants to reveal their experiences through three methods: walking interviews in which people living with dementia and occasionally a supporter took us on a 'neighbourhood walk' to show us around their local area; a filmed tour around their home; and a participatory social network mapping technique to explore relationships and social connections. In total, 127 people took part, completing 108 network maps, 100 walks and 59 home tours. All names reported here are pseudonyms.

Leaving the field in dementia studies

There has been considerable attention paid to designing research that can better engage with those living with dementia and enable them to document or record their own lived experiences (Novek and Wilkinson 2019), including adapting research methods to better 'fit' a dementia context and enabling those living with dementia to participate as fully as possible in research (Keady et al. 2017). Nonetheless, a recent review of dementia literature noted that 'relatively little has been written about how researchers and research practice must adapt *in preparation for and during data collection*, so that people living with dementia can be involved in research' (Webb et al. 2020: 2; our emphasis). So, while dementia studies may have started to pay attention to issues of research design and access to participants and entering the research field, the later stages of research, after data collection is completed, are largely absent from debate. That said, this is not to suggest that withdrawal from the field has been ignored in dementia studies. For instance, there has been discussion of risks around role confusion when boundaries between researcher and research participant might become blurred, and of the importance of 'signposting' participants who display signs of distress (McKeown et al. 2010; Novek and Wilkinson 2019).

Yet, such accounts arguably focus on the practical implications of withdrawing from the field, offering limited reflection of how leaving, and returning to, might influence knowledge production. Indeed, recent guidance from the UK Health Research Authority (which protects and promotes the interests of patients and the public in health and social care research) advises that:

> the definition of the end of the study should be documented in [research] protocol. In most cases, this will be the date of the last visit of the last participant or the completion of any follow-up monitoring and data collection described in the protocol. (HRA 2021)

This implies both a linear and a researcher-determined process of how and when withdrawal should happen, as well as a geographically and temporally fixed notion of where, when and what constitutes 'the field'. It also belies the ease with which leaving the field can be achieved, and offers a somewhat deterministic construction, bounded by visits to or interactions with participants, because designing research around a series of exits and re-entries, particularly with participants living with a fluctuating and complex cognitive and physical condition, can make locating end points difficult. In the remainder of this chapter, we explore what it means to leave and return through three broad issues: how we (re)engaged with participants at multiple re-entries to the field; supporting participants to return with us to the field; and the implications these might have for how the field is constructed in situations when participants might struggle to engage with, or easily recognise, the social worlds researchers are seeking to understand.

Remembering and being reminded of the field

Throughout the research we repeatedly returned to participants, sometimes in short succession such as after a few days, other times up to a year later, each time becoming reacquainted with the people and places we were researching. Each time, we were conscious that the lives we were returning to might be quite different to the ones we had been introduced to on earlier visits. This was compounded by an awareness that some participants living with dementia might not be able to easily remember taking part in the research or might have only vague recollections of doing so. Mindful that it might not be possible for some participants' memories to be 'jogged', we engaged in a process of 'active reminding' about their participation – a process that is common to other longitudinal research contexts. This included regular contact through postal and electronic mail communicating study updates so as to aid the continuation of engagement, though these could not be relied upon to always enable ongoing remembering. Where possible, we also worked with a friend or family member who could support this reminding, drawing on the trust between them and participants to secure ongoing involvement. This was not intended to manipulate the process of proxy consent to ensure ongoing involvement, but was about adopting a relational approach to participation in ways that we hoped were sensitive and appropriate.

Consequently, reminding was reliant on the relationship with each participant. We asked questions about their lives on topics they had previously shared, or might remind participants of the activity we had carried out together, such as the walk we had gone on, or met them at places we knew to be familiar. This active reminding helped to refamiliarise participants with us and our connection to them and, arguably supported the re-establishing of trust with participants as well as with family members and friends, through demonstrable familiarity. Crucially, this required being attentive to risks of assent or even coercion, as well as avoiding, as best we could, appearing to be over-familiar or even of 'faking friendship' (Duncombe and Jessop 2002).

Supporting participants to 'return to the field' consensually, especially in the absence of easy recollection, placed significant emphasis on the relationships we developed with participants, the knowledge we held about them and what we in turn did to maintain those relationships. This included deciding if, when and why it was appropriate to return to participants, and careful justification and articulation of our reasons for wanting to do so. It also placed importance on the processes and procedures we established to ensure these relationships remained fair and accommodating of changes in individual and well as relational dynamics.

Attempts to keep in touch through research updates, active reminding and requests to revisit participants may all be part of the toolbox of re-engagement but they are not neutral. For instance, when embarking on a return walk with Roger we reminded him of a previous interaction with us, repeating one of the stories he had shared:

> SC: We've been on one walk, haven't we? We walked down to the bowling club.
> Roger: Have we?
> SC: Yes. Yes, we did.
> Roger: Right.
> SC: And walked around so we'll do the same kind of walk again and just have a talk about … if anything's different since the last walk we went on.
> Roger: Yes. Yes.

Beyond a dementia contact, this might appear a banal and inconsequential effort of reminding, but in Roger's case we cannot be certain that he does remember our previous visits. Nor can we really know how he felt about our return and our efforts to initiate re-engagement with the research. However, this does not mean that Roger does not understand the purpose of this new exchange. Soon into the walk Roger asked again who the researcher worked for and where she lived as he attempted to place her and the context for our being with him. Roger was also confident in directing the walk and sharing stories about his local area, despite not necessarily remembering these previous encounters and requiring regular reminders about the purpose of our time together. There is an imbalance here, present perhaps in all longitudinal research (Miller 2015), but brought to the fore in the context of cognitive impairment, about when and to what extent we should re-share existing information with participants to support them to remember past encounters. In thinking through what sorts of information to re-share with participants, we are not seeking to deny people living with dementia the right to own and be reminded of their pasts but, rather, drawing attention to the ways and the contexts within which this might or might not be done. The reintroduction of 'snippets' of the participant's life through active reminding also reflected attempts on our part to redress some of the power imbalance between us and participants, presenting back to them in anticipation that they might regain ownership of those stories. On our return to conduct a home tour, Anna did not appear to remember our previous interactions, which had included visiting her at a dementia support group, going on a walk and a car journey and a trip to her local public house. We drew on these moments to invite Anna back to the field, arguably regardless of whether she wanted, or indeed was able, to offer any further comment on these previous interactions. In such ways, returning to the field was as much about facilitating participants' re-entries into their realm of experience as it was about our wanting to return to a point in participants' lives where we had previously left off.

Returning to different fields

For some participants, their dementia had changed to a point at which they struggled to participate in ways that we might have hoped. In such circumstances, the fields we were returning to were notably different socially, physically and cognitively, to those initially constructed. This is not to conflate the field with each participant, nor to make a point about difficulties in recognising how the field might have changed. Rather, it is to remark on how our intention to co-construct the field of experience for participants proved an ontologically, as much as analytically, daunting task because participants were no longer able to recognise the previous lives and experiences or how they might have changed.

For instance, Dennis had previously walked with us for over an hour and talked in depth about his experiences of living with dementia. In between research encounters, Dennis experienced a health setback which meant he could no longer manage taking part in the dementia activism which had been a central part of his life. Although he was able to take us on a walk, this was much shorter, terminating at the end of his street:

> Dennis: As I say, it's like ... you know, when I went out with the dog ... we'd go miles. But all that's gone. There's no point getting uppity about it because the reality is that's what it's got to be, you know. I would say as well, I'm walking down here and I'm just thinking ... [I've been] going up and down here for years. But now this all seems different to me.
> SC: Does it? It doesn't feel recognisable?
> Dennis: It's not the same, no. I still can see some of it ... but there always seems to be something different now than what it used to be. Especially if I wanted to go into [town] or anything like that, I wouldn't have a ... any chance at all.

For Dennis, the field has shifted in geographical and social scope and scale. There has been a reduction in the places he might be able to go to, or the contacts and relationships he can maintain. Although we were aware of these changes, we held back from pursuing them because of the potential to cause distress to Dennis that we had limited ability to redress, and which could potentially cause more harm to him than benefit for 'our' research.

Returning to other participants, we found some to be less mobile or not as sociable as they had been, though this was not necessarily always due to dementia. Celia, for example, had developed an illness *that* limited her physical abilities, which, combined with dementia made it difficult for her to walk and talk. She was no longer keen to participate in the research, and although her husband, Malcolm, continued to participate, the experiences and stories were no longer co-constructed, as they once had been, between the couple and the researcher. At one point, Malcolm shared his experiences of dealing with the changing situation:

> Lots of things can change, you know. I mean, ... I was getting no sleep at all and you lose all your focus. So eventually, you know, I put the bed down[stairs] ... just so I can get [some sleep]. It was when [Celia] was up all the time.

With some faltering, Malcolm described a period where Celia's health was precarious and had deteriorated significantly, and the impact this was having on him. The construction of the field here has shifted our view from one developed through our relationship with Celia, to that with Malcolm, and created a very different viewpoint from which we tried to understand something of Celia's experiences.

Changes in the symptoms of dementia mean that participants may no longer access the worlds that they used to, finding that their lives have become more restricting. This was emotionally challenging for participants to reflect upon, as well as for us to navigate. We wanted to ensure that we were 'doing the right thing' by maintaining, reducing or even ending participants' involvement. We were coming to understand about changes that, at times, some participants might struggle to identify, while being privy to private problems that in other circumstances might not have been revealed to a relative stranger outside the research context. This sense of invasion was particularly acute when participants had moved from the family home into care settings and when family members sometimes accounted for decisions in ways that inferred a desire a justify them. Some shared their sense of reaching 'breaking point' to explain why they were no longer able to support a loved one at home. Frank, for instance, described a decision about the care needs of partner Florrie:

> [The social worker] decided that I couldn't carry on the way I was. And he just said, 'it's just common sense, you're running yourself into the ground, and you can't keep going the way you are'. So he said, 'we're gonna have to do something about it' ... she was going to need permanent care. So, the social

worker, and the mental health visitor, between them, they decided that Florrie was going to need full time care.

Frank explained the changed circumstances that we found him and Florrie in, which were very different to those we had previously encountered. Indeed, our return visits to Celia, Malcolm, Frank and Florrie laid bare the emotionality of encounters and provoked consideration of whether participants might, or even could, continue to be involved in research in the context of deteriorating health. Returning to the field thus required assessing the extent to which individuals could participate in what might ostensibly be repeated phases of data collection but which were also processes that (re)created the field in emotionally complex ways. We sometimes got a sense of change ahead of any discussion, such as by visiting participants in new homes, but this did not offer much insight into the lived experiences we were seeking, aside from rather superficial observations of changed, and changing, scenes. It could also be difficult to facilitate ongoing engagement using the methods we hoped to deploy, especially if participants were no longer able to take us a on a walk outdoors or show us around their homes. For them, while we had hoped that our absence from the field might have been temporary, and despite their apparent eagerness for us to return, their situations were so significantly changed as to make previous fields out of our reach.

Extending the field

Our final set of reflections involve the ending of relationships with participants and our engagement with the field. At times this was stark, such as when participants became unwell or, for a small number, had died. Despite seemingly clear termination points, such participants do not simply vanish (Thorneycroft 2020). Relationships continued through our custodianship of, and analytical engagement with, their stories, and some participants returned to centre stage through the dissemination activities. Following our interactions for the purposes of data collection, a smaller number of participants joined us in dissemination activities, advising on activities, contributing to outputs and commenting on themes emerging from analysis. Their involvement in these stages extended the field in a somewhat different configuration. A group worked with us to develop several visual guides summarising the research, and we continued to have discussions about emerging findings, sharing of our analytical ideas and collectively reconstructing the fields we had all engaged in. We continued to encounter issues with memory, cognitive function and physical mobility which at times restricted engagement. Some participants stepped away from these activities because of changes to their symptoms. One participant had engaged in both phases of the research and was very active during early dissemination activities but decided to disengage from the work when travel and going out became more challenging. Her presence continued through the stories she had shared, such that her involvement was kept lively through an ongoing association with the research endeavour and she never fully left, at least not in our fieldwork imaginations. We are not suggesting that our experiences of engaging participants in dissemination here are novel, nor trying to reveal something new about where the field might be located. Rather, they force us to ask what it means, for participants and researchers, to leave the field indefinitely. For if the field is an intellectual endeavour, then, perhaps, we can never really identify the precise point at which we leave, any more than we can really locate the field that we are entering, returning to and leaving.

Discussion

Researchers can access plenty of advice about how to negotiate entry to and exit from the field and navigate relationships with others during research (e.g., Fox 2008; Gallaher 2011; LeCompte 2008). Oftentimes, accounts of leaving the field can seem somewhat unproblematic, or at least 'manageable', reduced to a case of getting in, getting data and getting out (Feldman et al. 2003). Yet the seemingly innocuous phrase of 'leaving the field' is a messy endeavour invoking much emotional toil (McGarrol 2017; Thorneycroft 2020). To leave the field implies, erroneously, a linear process at the end of which researchers and participants go back to their respective lives and carry on 'as normal', with the research act a temporary blip. Researchers and participants alike cannot simply put aside the emotions and histories that developed through previous interactions and then expect to 'pick up where they left off' at re-encounters.

The exits and re-entries we have described here are imperfect, not least because the field is as much a moveable idea as it is a location in time and space. So too is returning a relational act that it is not always obvious how to accomplish. In the context of dementia research, participants may have difficulties in recollecting previous interactions, even with support or prompting that meant our own exiting and returning was dependent on the layering of the relationships through each interaction. We acted and responded differently with different participants but remained imperfectly attuned to the situations to which we returned, not only because of the changes we might witness or be told about, but also because of the ongoing influence of these previous encounters and the status and form of the relationships with each participant that we had left off and perhaps naively hoped to simply pick back up. We worked hard to 'tread carefully' (Miller 2015), but still stumbled through the relationships and situations as we all, participants and researchers alike, tried to reacclimatise.

Leaving and returning to the field has inevitable implications for how data is collated, ordered and analysed to ensure its authentic recreation at the end of the research process. Used uncritically, they also suggest that researchers and participants can recognise when we are in the field, when we enter and leave it, and that researchers are adept at navigating the complexity of socio-spatial relations that constitute the field, in the quest for knowledge. The field though, is less a geographical and temporal location and more 'a set of relations nurtured, contested and developed during the course of several months' (Caretta and Cheptum 2017: 415). It emerges through the nexus of power in a collaborative, but not wholly equal, process of co-constructing (Gupta and Fergusson 1997). This is particularly apparent when it is researchers who drive the desire to re-engage with the field and, as in our case, may find themselves better equipped to drive its reconstruction.

Exiting and returning demands ethical attentiveness, most obviously perhaps around consent, but also to the avoidance of emotional discomfort or a candidness that might be too revealing (Duncombe and Jessop 2012). In the case of dementia, Hydén (2013) has argued for a shift of focus, in narratives, from their textual aspects as products to be created and analysed, to relational and embodied processes. Our return to the field creates a scaffold to encourage people living with dementia to tell new stories, reflect on past stories in new contexts and recast once-told tales in a different light. We re-engaged with this scaffolding, rather than with the specifics or 'realities' of what might or might not have changed in the intervening period between our visits, in order to remake relationships with participants, creating new moments of familiarity and connection, and ultimately recasting what, where and how we came to construct the field.

Conclusion

We are certainly not the first to worry about the issues we have discussed here. Others have agonised over how to avoid being (mis)interpreted as having exploited unequal relationships in a hunger to extract as much data possible from individuals considered by some to be vulnerable (Miller 2015). Nor are we alone in expressing caution in our engagement with those with whom we hope to establish longer-term research relationships. We find some reassurance in Oakley's (2016: 208) view of research materials as having been (cautiously) 'gifted' to us rather than 'somehow forced' into disclosure through faked friendships. Nonetheless, dementia adds another facet to how we might go about returning to the field and to what we might do with the stories bestowed on us while we are there. The negotiation of roles, interactions and relationships with participants, as well as emotions, ethics and expectations, are all important in the context of leaving and returning. Perhaps our experiences working in a dementia context bring these to the fore in ways that emphasise the complexity of what it means to depart from, return to and, ultimately, terminate the field. Consequently, we finish with consideration of three implications that emerge from our discussion.

First, it is worth restating that researchers should continue to recognise the field as more than a geographical and temporal location to be moved into and out of. It is constructed in moments of interaction with participants, through reflection on those interactions and during dissemination activities (Burrell 2009; Gupta and Ferguson 1997). Researchers are certainly familiar with reflecting on their own role in constructing the field, but perhaps should give a little more consideration to how they move on from the field in ways that remain sensitive to the experiences and intentions of research participants. This is not so much in terms of more careful articulation and negotiation of relationships but, rather, in recognising the different ways that participants may also leave the field and return to the field.

Second, leaving and returning to the field raises inevitable ethical implications. These include consideration of consent in cases when ongoing involvement may be assumed, or when participants are less able to consent to further engagement. Thought should be given to possible social and emotional attachment to the field for both participants and researchers (Duncombe and Jessop 2012). Returning to the field is an emotional process for both participants and researchers. Participants reveal experiences and views that they may subsequently not remember, in doing so creating dilemmas about what researchers should in turn re-share. Reminding, or not reminding, people of things they might have forgotten is a messy but unavoidable business. Each research engagement may start afresh, with the notable difference that we as researchers may well know more about participants than they are able to remember or than what they can make understood.

Finally, researchers should remain aware of the importance of renegotiating relationships and the ways in which the field itself is constructed through them. The field is no longer uncritically accepted as bounded spaces (Burrell 2009; Marcus 1995). As researchers engage and disengage with the field they are validating and narrating an experiential and cognitive rather than a physical movement (Amit 2000). Here, Rapport's (2000) focus on narrative awareness to clarify where and when to locate the field takes on a particular complexity in the context of dementia, though the wider point remains about how, and by whom, the field gets constructed. For leaving and re-entering the field is a relational act, dependent on interactional and cognitive as well as a physical and embodied conditions that are performed in moments of unequal understanding. Doing this alongside participants living with dementia might bring this into sharper focus, but it is arguably always part of how 'the field' gets understood and articulated.

Acknowledgements

This study was funded jointly by the Economic and Social Research Council (ESRC) and the National Institute for Health Research (NIHR). ESRC is part of UK Research and Innovation. The views expressed are those of the authors and not necessarily those of the ESRC, UKRI, NHS, the NIHR, or the Department of Health and Social Care. This work forms part of the ESRC/NIHR Neighbourhoods and Dementia mixed methods study (Reference number: ES/L001772/1). This chapter is based on reflections from the Greater Manchester fieldsite of Work Programme 4.

References

Alzheimer's Society (2021). 'Facts for the Media'. Available at https://www.alzheimers.org.uk/about-us/news-and-media/facts-media

Amit, V (ed.) (2000). *Constructing the Field: Ethnographic fieldwork in the contemporary world*. London: Routledge.

Burrell, J. (2009). 'The field site as a network: A strategy for locating ethnographic research'. *Field Methods*. 21(2): 181–199.

Caretta, M.A. and Cheptum, F.J. (2017). 'Leaving the field: (De-)linked lives of the researcher and research assistant'. *Area*. 49: 415–420.

Duncombe, J. and Jessop, J. (2012). '"Doing rapport" and the ethics of "faking friendship"', in T. Miller, M. Birch, M. Mauthner and J. Jessop (eds) *Ethics in Qualitative Research* (2nd edn). Pp. 108–121. London: Sage.

Feldman, M., Bell, J. and Berger, M.T. (2003). *Gaining Access: A Practical and Theoretical Guide for Qualitative Researchers*. Walnut Creek, CA: AltaMira Press

Fox, N. (2008). 'Leaving the field', in M. Given (editor) The SAGE Encyclopaedia of Qualitative Research Methods. Available at https://sk.sagepub.com/reference/research/n245.xml

Gallaher, C. (2011). 'Leaving the field', in Del Casino, V.J., Thomas, M.E., Cloke, P. and R. Panelli (eds) *A Companion to Social Geography*. Blackwell: Oxford, 181–97

Gupta, A. and Ferguson, J. (1997). 'Discipline and practice: "The field" as site, method, and location in anthropology', in G. Akhil and Ferguson, J (eds) *Anthropological Locations: Boundaries and Grounds of a Field Science*. Berkley: University of California Press, 1–46.

HRA [NHS Health Research Authority] (2021). 'Ending your project'. Updated 10 March 2021. Available at www.hra.nhs.uk/approvals-amendments/managing-your-approval/ending-your-project/

Hydén, L-C. (2013). 'Storytelling in dementia: Embodiment as a resource'. *Dementia.* 12(3): 359–367.
Keady, J., Hydén, L.C., Johnson, A. and Swarbrick, C. (eds) (2017). *Social Research Methods in Dementia Studies: Inclusion and Innovation.* Abingdon: Routledge.
LeCompte, M. (2008). 'Negotiating exit', in M. Given (ed.), *The SAGE Encyclopaedia of Qualitative Research Methods.* Available at https://sk.sagepub.com/reference/research/n284.xml
Marcus, G.E. (1995). 'Ethnography In/Of the World System: The Emergence of Multi-Sited Ethnography'. *Annual Review of Anthropology.* 24: 95–117.
McGarrol, S. (2017). 'The emotional challenges of conducting in-depth research into significant health issues in health geography: Reflections on emotional labour, fieldwork and life course'. *Area.* 49: 436–442.
McKeown, J., Clarke, A., Ingleton, C., Repper, J. (2010). 'Actively involving people with dementia in qualitative research'. *Journal of Clinical Nursing.* 19(13–14): 1935-1943.
Miller, T. (2015). 'Going back: "Stalking", talking and researcher responsibilities in qualitative longitudinal research'. *International Journal of Social Research Methodology.* 18(3): 293–305.
Novek, S. and Wilkinson, H. (2019). 'Safe and inclusive research practices for qualitative research involving people with dementia: A review of key issues and strategies'. *Dementia.* 18(3): 1042–1059.
Oakley, A. (2016). 'Interviewing women again: Power, time and the gift'. *Sociology.* 50(1): 195–213.
Rapport, N. (2000). 'The narrative as fieldwork technique: Processual ethnography for a world in motion', in V. Amit (ed.) *Constructing the Field: Ethnographic fieldwork in the contemporary world.* pp. 71–95. London: Routledge.
Thorneycroft, R. (2020). 'When does research end? The emotional labour of researching abjection'. *Methodological Innovations.* https://doi.org/10.1177/2059799120926350.
Ward, R., Clark, A., Campbell, S., Graham, B., Kullberg, A., Manji, K., Rummery, K. and Keady, J. (2018). 'The lived neighbourhood: Understanding how people with dementia engage with their local environment'. *International Psychogeriatrics.* 30(6): 867–880.
Webb, J., Williams, V., Gall, M. and Dowling, S. (2020). 'Misfitting the research process: shaping qualitative research "in the field" to fit people living with dementia'. *International Journal of Qualitative Methods.* 19. https://doi.org/10.1177/1609406919895926.

14

A constant apprenticeship in martial arts: the messy longitudinal dynamics of never leaving the field

David Calvey

This chapter is divided into four sections. The first outlines the context for this ethnographic project, including the biographical rationale and landscape. The second briefly explores martial arts as a distinctive field and my particular sociological gaze and lens on the topic. The third unpacks some confessional moments within my autoethnographic journey in martial arts. The fourth concludes the chapter and critically reflects on the lessons learned from this ethnographic tale of not leaving the field.

A biographical context and landscape: all roads led to Jeet Kune Do

Despite the clear opening up, democratisation and vast diversification of martial arts in modern society, away from the rather esoteric and closed communities of the past, the master–apprentice metaphor is still very pervasive with martial arts communities of practice. Deep loyalty to instructors, who often translate and interpret schools and styles of martial arts over dedicated life careers, mediate the journey of many martial art students. Family tree lineage is very influential here. Hence, the 'who taught you' question becomes a mark of status and longevity in many martial art circles. Hence, some students can follow particular instructors in a rather dogmatic fashion.

I moved around different styles from Shotokan Karate as a young teenager in a community centre to Aikido, Tai Chi, Jiu Jitsu, Kickboxing, Tae Kwon Do and Muay Thai, including private tuition in Thailand, in undergraduate and postgraduate university days.

I spent most of my time training in Jeet Kune Do (JKD) as a style, with a well-known and highly respected practitioner in that style in Manchester city centre. JKD is an eclectic style of martial arts famously originated by the creative and iconic Bruce Lee. As Jennings argues: 'it is a hybrid fighting system based on combat efficiency and street survival, as well as specific technical concepts such as broken rhythm' (Jennings 2019: 61). Although, at first sight, it seems simplistic and even crude as a martial art, it is a highly complex and innovative style, which favours repeated drills in natural stances rather than stricter pre-set Kata or forms. Indeed, it terms of learning JKD, our sensei (instructor/teacher) would continually implore us to see it as a method and not a style of fighting. As Bowman stresses:

> Lee clearly wanted to break away from *something* to do with conventional martial arts teaching and learning (specifically, the militarized, hierarchical, robotic, production-line approach, historically rooted in the first half of the 20th century, and its end results). (Bowman 2019: 36)

Part of the JKD I practised was specifically informed by Kali and Silat developments from Indonesia, which were very fine-tuned bodily manipulations and strikes. These technical issues were much more difficult to accomplish, different from anaerobic 'blasting the pads', with comparatively more simple kickboxing techniques.

My eclectic martial art past fitted in well with this flexible system of cross-training, which seemed less rule bound than other styles. Some have dismissed JKD as a legitimate martial art as a type of fake-fight choreography, due to Lee's legacy as a movie star. His fame in popularising to Western audiences what some have broadly referred to as *Kung Fu* has led, in part, to his marginalisation as a genuine martial art expert (Bowman 2010; 2013). Guro Dan Inosanto, who was a student of Bruce Lee, is recognised as a world leader in JKD, although JKD is not a completely unified system. Part of our ritualised homage to JKD was in attending a high-profile seminar given by Inosanto in Edinburgh. It was a sort of subcultural 'rite of passage' in drinking the night before and being able to competently do the international seminar the next day.

One particular aspect of the JKD philosophy was the disruption of hierarchy, displaced in the belt grading system typically found in most martial arts. This is a deeply symbolic measure of experience and rank that structures the shared belief system with the dan grade or black belt, the common inspirational end goal of many martial art students. In JKD belts are not worn, being associated with elitism, although grades are still taken. It has the advantage of flattening any sense of superiority and greatly encourages the idea of partnering up with a more diverse set of people, rather than being bounded by grade. It is a genuine acid test of your skills and continually gives immediate perspective on your personal progress.

This JKD philosophy is not commonly shared by other styles that I practised as I tried to change training partners and feel different energy flows, bodies and capabilities. I could learn from a range of different belts and not a set partner, which I found was practised less in other dojos (places of teaching). This contributed to my sense of constant apprenticeship.

Martial arts as a distinctive field of study

The distinctive study of martial arts is a relatively small and new academic field in the social sciences but a growing and highly creative arena. Part of the analytic push for studying martial art bodies is the wider and well-established complex debates on unpacking somatic and haptic phenomenology (Allen-Collinson 2011; Williams and Bendelow 1998), reflexive embodiment and body techniques (Crossley 2005; Shilling 2004; Turner 2008) and sensory experiences (Pink 2015; Sparkes 2017; Stoller 1989). Consequently, many of the key theoretical threads in the field have been broadly centred on embodiment (Spencer 2012; Channon and Jennings 2014) and habitus (García and Spencer 2013; Spencer 2009). Wacquant's (2004) study of boxing has been a pioneering milestone for many in this field.

The diasporic nature of martial arts has been clearly recognised (Delamont et al. 2017) in the field. A branch of the literature has also specifically explored the rise of mixed martial arts (Spencer 2009; 2012) and the increasing feminisation of martial arts (Channon and Matthews 2015), with others being concerned with the more traditional styles of Aikido (Foster 2015), Karate (Chamoli 2020) and Judo (Goodger and Goodger 1977). Much of the methodological push in the field is understandably in forms of autoethnography (Stenius and Dziwenka 2015), practitioner-researcher narratives (Jennings 2013) and autophenomenography (Allen-Collinson 2011).

Navigating a field that I could never leave

This section is organised around three broad conceptual categories and situated field vignettes.

Falsification, leadership and guruism

Much of the falsification was driven by guruism in the form of instructor and style loyalty and the desire to be subculturally accepted. One's attendance displayed a form of commitment, loyalty and allegiance to that sensei in that particular dojo. The question of doing other styles with different instructors was effectively blasphemy and heavily frowned upon. In particular, if you wanted to be graded to a dan grade or black belt, you had to demonstrate your senior membership by being a sort of believer, akin to a process of proselytisation in a religious faith (Lofland 1966). In some particular ways, martial arts devotion can be seen as pseudo religious (Jennings et al. 2010). Such guruism could result in hero-worship of the instructor, displayed in sycophantic and supplication behaviours in order to win favour and get closer attention in class. Guruism seemed to be more pronounced in styles which were associated with a martial art school or system set up privately by an individual rather than in large formal bodies like the British Judo Association and the Karate Union of Great Britain, where gradings were standardised and codified. Being graded by your own instructor clearly maintained guruism.

Hence, as I shifted around, despite being a brown belt in JKD, I had to continually accept and internalise a white belt novice status when in other styles. I was a constant apprentice in terms of my embodied learning in martial arts (Downey et al 2015; Wacquant 2005). Part of this involved me in obeying the rules of other styles when at their dojos, even when I zealously disagreed. I was regularly breaking etiquette rules and being an 'outsider' (Becker 1963) in different dojos. I particularly had to learn to supress 'broken rhythm' principles when training elsewhere. For example, when I included an elbow in an Aikido class technique this was frowned upon, much as when I included a reap and throw in a Tae Kwon Do class. In these settings I had to effectively repress my bodily expertise and follow their strict instruction, without deviation. I understood and appreciated style discipline, but still found it very frustrating.

In the classic Weberian sense (Weber 1921), the instructor was a charismatic leader, whose legitimacy as a basis of authority was expert knowledge, clear lineage and honorific esteem as a martial art practitioner. In my case my chief instructor was a 9th Dan in Shukokai Karate and certified international instructor in Muay Thai, JKD, Kali and Silat. Related to this, the role of the follower is an active rather than passive one. As Hollander argues in his relational view of leadership, this vitally 'depends upon the followers' recognizing the leader's special attribute' (Hollander 1992: 52). Despite the diversity and democratisation of modern martial arts, guruism is widespread in such communities of practice (Wenger 1998).

An interesting demonstration of 'following' occurred one night in the JKD dojo when, in free sparring, which is usually restrained, a senior black belt was very heavy in his full contact with me, without prior consent. I genuinely thought I was going to be injured, but fortunately survived the bruising encounter. The duration seemed like hours rather than minutes, due to the intensity. The sensei never intervened in any way, and I was eagerly waiting for his usual reminders of 'take it easy', 'tone it down' and 'steady' in a stern, raised tone. I read the encounter as a pre-black belt grade ritual, and also an endorsement of following his style as real-world self-defence and authentic street combat. A black belt grade also had instructor status and implications. It felt like a mark of obedience. I never discussed it with the instructor, as if it was something I had to endure and not question in any way. I did not see this as a form of deviant cultish conversion (Lofland

and Stark 1965), but more sensibly as the 'social logic' of the subculture (Wacquant 1992).

Managing the disconnect and divided self between covert bouncing and martial art training

I was conducting my longitudinal study of bouncing at the same time as training in martial arts, which had certain obvious advantages of bodily craft (Wacquant 2004) and physical capital (Shilling 2004). The problem was that, despite many of my martial art peers having both done some bouncing in the past, with some others still active in the profession, I kept it a secret from them, as my study was purely covert and consistently hidden from my fellow bouncers.

The fictive kinship (Woodward and Jenkings 2011) and sense of shared camaraderie and community solidarity was very pronounced in the bouncing community, hence my covert academic lens would have been frowned upon, which in turn would have caused distance and distrust with my martial art colleagues. Thus, when we were discussing an application of technique in a sensible 'real-world' conflict scenario, it would often be drawn from the many years of bouncing that our chief instructor had done. I would repress my urge to enthusiastically confirm this with my own experiences of bouncing, but instead feign surprise, deference and wonderment. This managed memory repression continued over a lengthy period and was a source of guilt for me.

A parallel example of this occurred while bouncing. I had to intervene in a brawl in a bar in Manchester city centre, which at the time was a hotspot for gangs. One aggressive young man had 'glassed' a stranger and I had to quickly restrain him before help arrived from my fellow bouncers. This happened in the basement dance-floor section of the busy bar, where I was on patrol alone, while two other bouncers fronted the entrance door. Typically, there would have been two bouncers downstairs, but the second bouncer was running late that evening.

I managed to knock the glass out of the man's hand by striking his wrist and then wrapping his arms tightly against his body in a type of trapping action. I was then able to pin him against the nearest wall to immobilise him. Adrenalin was pumping and I was very fearful of immediate retaliation by his friends in the bar. The bar staff were overworked and the place was awash with bottles and glasses, so I was acutely aware of the risk of being hit over the head from behind. The other bouncers assisted, and the basement bar was quickly cleared. The bar staff told the bouncers what had happened, and they were keen to know where I trained, so they could take some classes with me. This was a sign of camaraderie and respect for me, but I deliberately lied about where the dojo was, stressing that it was a distant and gruelling trek. The truth, ironically, was that it was close by in the city centre.

A few fieldwork close shaves also occurred. The first was when a martial art friend suggested that on a Thursday evening after training they would go to a certain bar in town where I was doing a bouncing shift. I assertively put them off, saying the drinks were very expensive and the clientele were rude and snooty, which was far from the truth. On another occasion a female bouncer with whom I worked was the former wife of a senior martial artist from the dojo, who was at associate instructor level. Fortunately for me, she was estranged from him and never discussed it.

I envisaged my covert bouncing role being awkwardly unravelled, as some of my martial art peers knew that I was a university sociology lecturer, but no more detail than that; and my bouncing peers knew that I did martial art but not that I was affiliated with a university. It was about managing a partial knowledge mosaic in both groups, and a deliberate separation of networks. My work and leisure spheres began to merge for sociological study.

A good example of liminality and managed deception was my relationship to Greg, a person I knew who was the same age as me and was a senior student at the JKD dojo. He was both a veteran bouncer and an accomplished martial artist, and now runs his own martial arts dojo. He knew that I was doing doors and we had some adventures when I saw him around, in terms of free entry and free drinks. Our relationship was one of mutual adoration and affectionate rapport, often displayed in the hyper-masculine horseplay of trying out martial art 'moves' and sharing humorous war stories and door banter, but he never knew that I was doing a covert study of bouncing. Although he connected me to the university, he assumed that I was a mature student and that bouncing was a means of earning some cash to support myself.

While this was a form of fake friendship (Duncombe and Jessop 2012), it was, ironically, genuine – although the risk (Lee-Treweek and Linkogle 2000) of it souring was ever present. I was ultimately involved in a type of edgework (Lyng 1990). Fieldwork roles, unlike Gold's (1958) classic typology, are much more fluid and paradoxical. Managing these secrets became a normalised part of my fieldwork and 'divided self', although not in any traumatic sense. It was a 'hidden struggle' (Punch 2012) within my fieldwork.

The spoiling effects on a life-long hobby

I had practised martial art from my early teens in various forms and styles. As my interest in carnal, bodily and corporeal sociology increased and deepened at the same time, I found myself slowly dissecting and analysing my martial art journey. My field of play slowly but surely became my field of study.

My identity as a martial art student was also tied up with my temporal and shifting bouncing identity. It was a complex web of deceit, white lies and frank intimacy, as my bouncing research was covert and continued for a lengthy period of time. One of the most significant aspects of the spoiling effects on my martial art practice was that I would often stay in my head rather than my body. Namely, I would over-analyse and over-theorise a bodily move or sequence rather than drill it, which affected my abilities and growth as a martial artist. Do it rather than over think it was a familiar thought. So, rather than getting better at a neck choke, I would be immediately thinking of a sensory ethnography essay on it.

Another related spoiling effect was that I was studying myself as a martial artist, and at times my vulnerable self. Looking in the mirror, although credible in terms of autoethnography, was also tiring and demanding. It became a narcissistic echo-chamber. What different people know of me became blurred over time as my martial art, academic and bouncer selves converged.

Because of the guilt attached to the deceit in some of these versions of the self, one of the most significant spoiling effects was on the mediation of friendship. I made friendships throughout my martial arts journey, but some were on faked grounds. I could not tell those people that they were also objects of academic study. I continually used impression-management tactics to navigate and manage my guilt. I had developed what Watts (2008) describes as 'attachments' in her discussion of emotion, empathy and exit in her ethnography of a cancer drop-in centre.

Rossing and Scott (2016) characterise this ambivalent state as 'taking the fun out of it' in their studies of aerobics and swimming. They stress:

> Doing activity-based ethnography on something personally special is a double-edged sword: on the one hand elucidating awareness, but on the other depriving the researcher of pleasure and 'spoiling' aspects of their identity.

Like Fincham (2008) in his ethnography of bicycle messengers, and seminally like Becker and his jazz musicians (1951), work and leisure distinctions in my study were not kept binary but become blurred and messy. Mac Giollabhui et al's (2016) ethnography of covert policing has some resonance for me as regards the researcher's participation in tests of allegiance and the building of trust. Ultimately, getting an insider view of my martial art journey had a 'spoiling effect' on my hobbyist field, contributing to forms of 'ontological destabilization' and identity blurring.

Some lessons learned

The methodological reflections from my case study point towards the saturated nature of emotionality in some field research journeys. I have still not told my JKD sensei about my covert bouncing journey, despite him being an experienced bouncer himself. My assumption is that he would approve of my being a bouncer and practically applying and testing my martial art corpus of skills, but not of the deceptive nature of the inquiry. It would sully our relationship, hence I have still kept it a secret. His affectionate reference to me as 'Dr D' is a positive memory that I am very attached to. It has been a protracted process of managing ambivalence in the field, which continues today.

I have experienced my martial art journey very differently from my peers, by never fully turning off the 'sociological analysis machine' about embodiment and habitus. My analytic dream was in resisting sentimentality (Becker 1967) and not turning my martial art peers into yet more academic forms of exotica, which can ultimately lead to the 'disappearance of the phenomena' (Garfinkel 1967; 2006) and types of academic zoo keeping (Gouldner 1968).

More generally, leaving out and correctively glossing absence, loss and mistakes in field research (Smith and Delamont 2019) overly sanitises research methodology and removes its colourful warts and multilayered mosaic character. I have never left the field and still practise martial art, although my covert study of bouncing has finished. I envisage that I would never get any form of emotional closure in that I don't feel that I could tell either the bouncers or martial art peers the full story, as it seemed a dual betrayal. The bonds and friendships I had developed, although genuine in some respects, were faked in other ways. This was compounded and intensified as I would occasionally encounter bouncers that I knew when I had finished the study, and I had to get 'back into character' in my post-fieldwork biography (Calvey 2017; 2019; 2000; 2021).

The longitudinal nature of my immersion bled into my identity politics. Taylor reflects on the paradoxical nature of leaving the field, stressing: 'so sometimes leaving the field means staying in the field and struggling with the human issues raised by the fieldwork' (Taylor 1991: 247). Thus, positionality was a complex process of 'being on both sides' (Calvey 2021). My commitment to empathy in field relationships was tied up with guilt. As Gable claims, 'Guilt emerges out of moral mutuality. Moral mutuality exemplifies ethnographic intersubjectivity' (Gable 2014: 256). My 'heartful' autoethnography (Ellis 1999) was saturated in moral compass conundrums and emotionality (Carter and Delamont 1996; Holmes 2010). I recognise and appreciate both the diversity of autoethnography (Tolich 2010) and its narcissistic overtones and narrative blind alleys (Atkinson 1997; Delamont 2009), but it was methodologically appropriate in these projects.

These reflexive sketches of my lived experience and impressions of being a martial artist are attempts at faithful representations of the field rather than prescriptive recipes or naive manifesto calls. In my case, I never fully left the field, but managed my particular odyssey within it. I have excavated my narrative as a 'layered account' (Ronai 1995) in a spirit of candour, fragility and authenticity, rather than as another caricatured, vain and heroic fieldwork trope.

The sentiments and sensibilities here are to recognise, appreciate and fully embrace the dynamic that ethnographic exits are not merely crude and reductionist techniques or a fixed fieldwork stage but complex, liminal and ambivalent endeavours, which includes types of lengthy revisiting. The assumption that managing rapport in fieldwork entries is often typically prioritised over relatively easy exits has to be challenged. The methodological literatures seem utterly biased towards the former.

What to some are apparently failed exists or ethnographic mistakes, on closer inspection are extended and emotional fieldwork immersions. The simple and naive duality of fieldwork entry and exit does not adequately reflect the messy nature of ethnography. Such ambiguity, liminality and complexity around ethnographic exits should be accepted, indeed celebrated, aspects of the creative and imaginative craft of ethnography (Atkinson 2015) and not sanitised out as methodological horrors. In some ethnographies, 'should I stay or should I go' still endures as a very puzzling question and reality.

References

Allen-Collinson, J. (2011). 'Intention and epochē in tension: Autophenomenography, bracketing and a novel approach to researching sporting embodiment'. *Qualitative Research in Sport, Exercise and Health*. 3(1): 48–62.

Atkinson, P.A. (1997). 'Narrative turn or blind alley?' *Qualitative Health Research*. 7(3): 325–344.

Atkinson, P.A. (2015). *For Ethnography*. London: Sage.

Becker, H. (1951). 'The professional dance musician and his audience'. *American Journal of Sociology*. 57(2): 136–44.

Becker, H.S. (1963). *Outsiders: Studies in the Sociology of Deviance*. New York: Free Press.

Becker, H.S. (1967). 'Whose side are we on?' *Social Problems*. 14(3): 239–247.

Bowman, P. (2010). *Theorizing Bruce Lee: Film-fantasy-fighting-philosophy*. Amsterdam: Rodopi.

Bowman, P. (2013). *Beyond Bruce Lee: Chasing the Dragon through Film, Philosophy and Popular Culture*. New York: Columbia University Press.

Bowman, P. (2019). 'Fighting over Bruce Lee'. *Martial Art Studies*. 8: 29–48.

Calvey, D. (2017). *Covert Research: The Art, Politics and Ethics of Undercover Fieldwork*. London: Sage.

Calvey, D. (2019). 'The everyday world of bouncers: A rehabilitated role for covert ethnography'. *Qualitative Research*. 19(3): 247–262.

Calvey, D. (2000). 'Getting on the door and staying there: A covert participant observational study of bouncers', in G. Lee-Treweek and S. Linkgole (eds) *Danger in the Field: Risk and Ethics in Social Research*. pp. 43–60. London: Routledge.

Calvey, D. (2021). 'Being on both sides: Covert ethnography and partisanship with bouncers in the night-time economy'. *Journal of Organizational Ethnography*. 10(1): 50–64.

Carter, K. and Delamont, S. (eds) (1996) *Qualitative Research: The Emotional Dimension*. Aldershot: Avebury.

Chamoli, A. (2020). 'The untold warriors of Shotokan Karate (an introduction to Shotokan) regarding India and Japan'. *International Journal of Science and Research*. 9(12): 1251–1253.

Channon, A. and Jennings, G. (2014). 'Exploring embodiment through martial arts and combat sports: A review of empirical research'. *Sport in Society*. 17(6): 773–789.

Channon, A. and Matthews, C. (eds) (2015). *Women Warriors: International Perspectives on Women in Combat Sports*. Basingstoke: Palgrave MacMillan.

Crossley, N. (2005). 'Mapping reflexive body techniques'. *Body and Society*. 11(1): 1–35.

Delamont, S. (2009). 'The only honest thing: Autoethnography, reflexivity and small crises in fieldwork'. *Ethnography and Education*. 4(1): 51–63.

Delamont, S., Stephens, N. and Campos Rosario, C. (2017). *Embodying Brazil: An Ethnography of Diasporic Capoeira*. London: Routledge.

Downey, G., Dalidowicz, M. and Mason, P.H. (2015). 'Apprenticeship as method: Embodied learning in ethnographic practice'. *Qualitative Research*. 15(2): 183–200.

Duncombe, J. and Jessop, J. (2012). '"Doing rapport" and the ethics of "faking friendship"', in T. Miller, M. Birch, M. Mauthner and J. Jessop (eds) *Ethics in Qualitative Research*. pp. 108–121. London: Sage.

Ellis, C. (1999). 'Heartful autoethnography'. *Qualitative Health Research*. 9(5): 669–683.

Fincham, B. (2008). 'Balance is everything: Bicycle messengers, work and leisure'. *Sociology*. 42(4): 618–634.

Foster, D. (2015). 'Fighters who don't fight: The case of aikido and somatic metaphorism'. *Qualitative Sociology*. 38: 165–183.

Gable, E. (2014). 'The anthropology of guilt and rapport: Moral mutuality in ethnographic fieldwork'. *HAU: Journal of Ethnographic Theory*. 4(1): 237–258.

García, R.S. and Spencer, D. (eds) (2013). *Fighting Scholars: Habitus and Ethnographies of Martial Arts and Combat Sports*. London: Anthem Press.

Garfinkel, H. (1967). *Studies in Ethnomethodology*. Englewood Cliffs, NJ: Prentice Hall.

Garfinkel, H. (2006). *Seeing Sociologically*. Boulder, CO: Paradigm Publishers.

Gold, R.L. (1958). 'Roles in sociological field observations'. *Social Forces*. 36(3): 217–223.
Goodger, B.C. and Goodger, J.M. (1977). 'Judo in the light of theory and sociological research'. *International Review of Sport Sociology*. 12(2): 5–34.
Gouldner, A.W. (1968). 'The sociologist as partisan: Sociology and the welfare state'. *The American Sociologist*. 3(2): 103–116.
Hollander, E.P. (1992). 'Leadership, followership, self, and other'. *Leadership Quarterly*. 3(1): 43–54.
Holmes, M. (2010). 'The emotionalization of reflexivity'. *Sociology*. 44(1): 139–154.
Jennings, G. (2013). 'Interviews as embodied interaction: Confessions from a practitioner-researcher of martial arts'. *Qualitative Methods in Psychology Bulletin*. 16: 16–24.
Jennings, G. (2019). 'Bruce Lee and the invention of Jeet Kune Do: The theory of martial creation'. *Martial Art Studies*. 8: 60–72.
Jennings, G., Brown, D. and Sparkes, A.C. (2010). 'It can be a religion if you want: Wing Chun Kung Fu as a secular religion'. *Ethnography*. 11(4): 533–557.
Lee-Treweek, G. and Linkogle, S. (eds) (2000) *Danger in the Field: Risk and Ethics in Social Research*. London: Routledge.
Lofland, J. (1966). *Doomsday Cult: A Study of Conversion, Proselytization and Maintenance of Faith*. Prentice Hall: New York.
Lofland, J. and Stark, R. (1965). 'Becoming a world-saver: A theory of conversion to a deviant perspective'. *American Sociological Review*. 30(6): 862–875.
Lyng, S. (1990). 'Edgework: A social psychological analysis of voluntary risk taking'. *American Journal of Sociology*. 95(4): 851–886.
Mac Giollabhui, S., Goold, B. and Loftus, B. (2016). 'Watching the watchers: Conducting ethnographic research on covert police investigation in the United Kingdom'. *Qualitative Research*. 16(6): 630–645.
Pink, S. (2015). *Doing Sensory Ethnography*. (2nd edn). London: Sage.
Punch, S. (2012). 'Hidden struggles of fieldwork: Exploring the role and use of field diaries'. *Emotion, Space and Society*. 5(2): 86–93.
Ronai, C.R. (1995). 'Multiple reflections of childhood sex abuse: An argument for a layered account'. *Journal of Contemporary Ethnography*. 23 (4): 395–426.
Rossing, H. and Scott, S. (2016). 'Taking the fun out of it: The spoiling effects of researching something you love'. *Qualitative Research*. 16(6): 615–629.
Shilling, C. (2004). 'Physical capital and situated action: A new direction for corporeal sociology'. *British Journal of Sociology of Education*. 25(4): 473–487.
Smith, R.J. and Delamont, S. (eds) (2019). *The Lost Ethnographies: Methodological Insights from Projects That Never Were*. Bingley: Emerald.
Sparkes, A.C. (ed.) (2017). *Seeking the Senses in Sport and Physical Culture: Sensuous Scholarship in Action*. London: Routledge.
Spencer, D.C. (2009). 'Habit(us), body techniques and body callusing: An ethnography of mixed martial arts'. *Body & Society*. 15(4): 119–143.
Spencer, D.C. (2012). *Ultimate Fighting and Embodiment*. London: Routledge.
Stenius, M. and Dziwenka, R. (2015). 'Just be natural with your body: An autoethnography of violence and pain in mixed martial arts'. *International Journal of Martial Arts*. 1: 1–24.

Stoller P. (1989). *The Taste of Ethnographic Things: The Senses in Anthropology*. Philadelphia, PA: University of Pennsylvania Press.
Taylor, S.J. (1991). 'Leaving the field: Research, relationships and responsibilities', in W.B. Shaffir and R.A. Stebbins (eds) *Experiencing Fieldwork: An inside View of Qualitative Research*. pp. 238–247. London: Sage.
Tolich, M. (2010). 'A critique of current practice: Ten foundational guidelines for autoethnographers'. *Qualitative Health Research*. 20(12): 1599–1610.
Turner, B. (2008). *The Body and Society: Explorations in Social Theory*. (3rd edn). London: SAGE.
Wacquant, L. (1992). 'The social logic of boxing in Black Chicago: Toward a sociology of pugilism'. *Sociology of Sport Journal*. 9(3): 221–254.
Wacquant, L. (2004). *Body and Soul: Notebooks of an Apprentice Boxer*. Oxford: Oxford University Press.
Wacquant, L. (2005). 'Carnal connections: On embodiment, apprenticeship, and membership'. *Qualitative Sociology*. 28(4): 445–474.
Watts, J.H. (2008). 'Emotion, empathy and exit: Reflections on doing ethnographic qualitative research on sensitive topics'. *Medical Sociology Online*. 3(2): 3–14.
Weber, M. (1921). 'The sociology of charismatic authority'. Republished in translation (1946) in H.H. Gerth and C.W. Mills (trans. and eds), *From Max Weber: Essays in Sociology*. pp. 245–252. New York: Oxford University Press.
Wenger, E. (1998). *Communities of Practice: Learning, Meaning and Identity*. Cambridge: Cambridge University Press.
Williams, S. and Bendelow, G. (1998). *The 'Lived' Body: Sociological Themes, Embodied Issues*. London: Routledge.
Woodward, R. and Jenkings, K.N. (2011). 'Military identities in the situated accounts of British military personnel'. *Sociology*. 45(2): 252–268.

Part IV

Returns, responsibilities and representations after 'leaving'

15

A cautionary tale about 'respondent validation': the dissonant meeting of 'field self' and 'author self'

Daniel Burrows

Introduction

In this chapter I reflect on my experiences of respondent validation following a publisher's acceptance of a book proposal based on ethnographic fieldwork I had previously conducted with a social work team. I explore how returning to the field as an author can give rise to a renegotiation of field relations, since it involves the presentation of an 'author self' in place of a former 'field self'. I argue that engaging in respondent validation means the continuation of field relations despite the shift in the self presented by the researcher, and recount how discomfort can arise for both researcher and participant if this is overlooked.

It is not unusual, and it is usually not uncomfortable, for individuals to know and inhabit multiple versions of self. In most social situations, social actors can adjust their behaviour and expectations of others in a way that is appropriate for the social frame in which they are placed, helping them to create a shared sense of meaning (Goffman 1974). Thus, an individual who is part of a sports team will know the kind of language and gestures, and topics of conversation, to expect and employ around teammates, yet would be comfortable presenting and interpreting an entirely different set of social cues around, say, their family members or work colleagues. Our ability to adjust our outward behaviours and to interpret others' behaviours depending on social context reflects the fluidity and multiplicity of the self. The 'me' on which the 'I' reflects (Mead 2011 [1913]) does not only respond to a 'general other', but to the other(s) involved in the particular social situation. When we are involved in interaction, the presentation of ourselves is for a specific audience, meaning that some aspects of identity will be emphasised and others hidden; that some emotional responses might be repressed or exaggerated; and that our physical form will be presented in a particular way (e.g., our posture, our distance from the other person, our gestures). Discomfort can arise, however, if an individual finds that they are negotiating two frames simultaneously, since the self-presentations we have in those different frames may not be easily reconcilable. For example, if a lecturer and a student were to come across each other at a party outside the university, both might struggle to know how to react, since the enactment of the student–teacher relationship, with its power relations and expectations on both sides of formality, might be inappropriate at an occasion devoted to frivolity and merriment. Such occasions highlight to social actors the contrasting roles they play and can give rise to a sense of their own hypocrisy, since they are made aware of the inconsistencies in their own behaviour between different social settings. The charge of hypocrisy in such circumstances is often an unfair one, however, since the 'I' can act spontaneously, according to the individual's conscience and inclinations, even while the 'me' is adjusted according to the social situation.

While the adjustment of our social selves is often unconscious or given only the barest thought, doing ethnography requires us to think carefully about the selves we inhabit. We must be conscious of at least two distinct selves – the 'field self' responsible for collecting data, and the 'author self' involved in analysing the data gathered and presenting their findings as scholarship. Whereas the 'author self' emerges largely through private contemplation, the 'field self' is produced through live interaction and needs careful crafting, since impression management is central to fieldwork (Hammersley and Atkinson 2019). Impression management enables us to negotiate access to the field at the beginning and ensures that our presence continues to be legitimate throughout the process of data collection. Of course, our self-presentation must evolve and adjust as our work progresses – for example, there are some questions that are appropriate to ask only when we are seen as newcomers within the field and others that would not be appropriate to explore until we have established a strong rapport and a decent level of trust (Delamont and Atkinson 2021). Yet, while we must think deeply about and work hard at maintaining field relations (Coffey 1999), and while successful fieldwork necessarily involves some level of secrecy about the nature of our work as we gather data (Roth 1962; Fine 1980), it is impossible to do ethnography without the enactment of an authentic encounter between the researcher and the field. Learning through ethnography happens because of the encounter between the self and the other – a moment of unknowing and striving to learn the perspectives of the other (Hickey and Smith 2020). In Mead's terms, the 'I', whose responses are spontaneous and incalculable, must be present along with the crafted 'me'. The parallel existence of an 'author self' in no way should be thought to diminish the importance of an authentic self in the field.

The fieldwork I undertook was carried out with a team of hospital social workers in the UK, in fulfilment of the thesis component of a professional doctorate. From the outset, I did not find it a challenge to craft field relations while being authentic in my interactions with participants. As a former social worker, I was alert and sympathetic to the challenges of the work in this setting, which focusses on arranging for the discharge of older people from hospital in a context of high pressure from hospital staff and managers to clear beds, while balancing the complex needs of those people and their families (see Burrows 2020). In negotiating access to the field, I emphasised to the social workers that I was one of them, and that I regarded their work as an unsung specialty[1] that was particularly worthy of study. I enjoyed each day of fieldwork, valued getting to know each participant and was happy to share aspects of my own life with others, when it felt appropriate to do so. I was one of only three men in the field, with the majority of social workers being women, a situation that I was used to and comfortable in from my previous career.

While forming and maintaining cordial social relationships in the field reflected parts of my personality that remain roughly the same no matter the social context in which I find myself, I did become conscious of an aspect of my 'field self' that was specific to the social context of doing ethnographic research: that of the 'socially acceptable incompetent' (Hammersley and Atkinson 2019). The types of situations, tasks and dilemmas confronting these social workers every day were vastly different to those I had encountered in my own practice as a social worker, and there were unfamiliar acronyms and jargon words to learn. Doing ethnography depends not only on being able to see and describe what is in front of you, but on being able to interact with people in the field so that they can show you aspects of their world and its constitution (Gobo and Marciniak 2016). Inhabiting this innocent, almost naive, self was unsettling to me because, in contrast to the social workers who were daily involved in making decisions and getting things done, I possessed very little agency – it was not my place to intervene, advise or even share an opinion, but simply to bear witness, record and learn. I felt, during the fieldwork, as though I were a benign, inert presence around the setting; an occasional (hopefully) pleasant distraction, sometimes a sounding board for thinking about dilemmas or conflicts, but ultimately a harmless, genial onlooker.

The time I spent in the field (which totalled thirty working days spread over a period of four months – an admittedly short period for an ethnographic study) passed without any difficulties worth remarking upon. My decision to leave the field was heavily influenced by time limitations, but it was also true that I already felt I had reached theoretical saturation in some key thematic areas. There had been an implied agreement when I negotiated access that I would share my analysis of my findings with the social workers, and therefore I made a conscious decision to keep in touch and to arrange to return to the team on occasions to discuss my findings. The team were happy for this to occur, and over the course of the next couple of years I arranged meetings on a few occasions to come into team meetings and talk about what I had been writing about. I treated these meetings as a form of respondent validation, because I wanted to find out the extent to which my interpretations of the setting rang true to the social workers. Though I always felt mildly nervous before visiting with the team to talk about my findings, the meetings we had were consistently friendly and supportive. I had come away from the field with a favourable impression of the social workers, who struck me as hard working, thoughtful and conscientious. Much of what I found to be of interest in my findings was the contrast between the heavily bureaucratised nature of the work the team were doing on a daily basis and the high level of commitment they showed to the fundamental social work values of human rights, social justice and empowerment (see Burrows 2020). As such, I was critical of the systems within which the social workers practised, over which they had little control, but highlighted the ways in which they enacted their values and showed high levels of skill in their dealings with patients, family members and clinicians. Usually, when I spoke to the team, I did so in general terms, often not referring in detail to individual fieldnotes or interview excerpts. Thus, I was able to palliate any remarks about their practices that might otherwise have felt disparaging, and to stress that any criticism I had was for the faceless 'bureaucratic system', and not for them as individuals.

Contact with the team naturally tailed off after I submitted my thesis. I presented them with a copy, of course, and thanked them sincerely for their involvement in the study. Sometime later, however, I found that my proposal to convert the thesis to a book had been accepted by a publisher, and happily shared this news with the team via an e-mail, proposing that we should meet up again to discuss their ideas about future priorities in the development of hospital social work. It took a few weeks before a slot was available, and when the meeting eventually took place I could sense a new tension. At the start of the meeting, one team member told me that another participant, who was not able to present, had said to 'ask Dan why he made us look like such a bunch of muppets'.[2] No explanation was forthcoming as to where or how my work had done this, and overall the meeting still felt positive. A few days afterwards, however, the gatekeeper with whom I had originally begun access negotiations wrote to inform me that there was growing disquiet in the team about the way they felt that snapshots of their practices had been taken out of context. Over a short chain of e-mails this hardened into an expression of concern on several individuals' part that the findings made the team look unprofessional, and the question was raised as to whether I would be entitled to publish the book without consent. I was even asked to send a copy of the consent form one person had signed, as it was intimated that it was never agreed with participants that the work could be published beyond the submission of a thesis.

This exchange of e-mails left me profoundly shaken. I questioned whether I had acted ethically towards participants in my access negotiations, field relations and writing up.[3] Checking my consent sheets confirmed that publication had been discussed, and, as far as I could remember, I had always been open in conversations with participants about my ambitions to publish after the thesis was completed. Senior colleagues assured me that I owned the copyright on my data and was entitled to publish whatever I saw fit. These considerations provided me with little comfort because, whatever legal rights I had, my personal integrity in my relations with the team remained in question. The good opinion and good wishes of each member of the team were important to me and it was painful to imagine that their previously warm and cordial disposition towards me had been replaced with suspicion and resentment.

Clearly, an ethnographer must reserve the right to describe and analyse the world as they find it, and I did not feel duty bound to write only in praise of the team; but I could not countenance the idea of publishing the book in knowing defiance of their objections. I do not hold it to be a general principle that ethnographers should always seek the approval of their participants for the work they produce, and in many cases there may be good reason not to. My sense of obligation towards my participants in this regard, however, was shaped particularly by the manner in which I had chosen to leave the field. By agreeing to visit the team to provide updates on my analysis and seeking their feedback on my developing ideas, and by inviting them to contribute some thoughts about how hospital social work might be shaped in the future, I had implied a relationship between us in which, as an author, I would seek approval of my work. Effectively, this meant that the complex negotiation of consent that is an ongoing feature of fieldwork relations in organisational contexts (Plankey-Videla 2012) was extended into writing up and publication. I had not anticipated this when I adopted 'respondent validation' as a means of enhancing the credibility of my work (Guba and Lincoln 1994), and it did not occur to me when revisiting the field that consent negotiation was still ongoing.

Had I realised that my continued contact with the team had these implications, I might have taken more time to discuss the parts of my work with individuals in which there was critical analysis of their practices. During the process of revisiting the field and discussing my ideas, my writing had not been scrutinised by the participants and I did not press for this scrutiny. However, if the negotiation of consent is ongoing through the writing-up and publication process, then it is vital that participants understand what they are consenting to. My verbal presentation of the parts of my work in which there was implied criticism of aspects of the social workers' practices was designed to minimise social awkwardness and to reassure the participants of my high regard for their work. As I found out, ethnographers need to resist such impulses when seeking respondent validation, and to think carefully about ongoing field relations.

When I told the participants in my study that I was intending to publish my work as a book, I invited the kind of scrutiny from them as readers that, hitherto, I had avoided. The 'shallow cover' (Fine 1980: 123) under which I had carried out the fieldwork was gone. While they perceived that my research was only for a thesis, the participants did not appear too worried about the impression they might make on the limited number of people who would read it. When faced with the prospect of a published book, however, the need to consider the impression that their doings and sayings might make on a reader became much more pressing. It is my understanding, therefore, that concern over the impression they would be making led participants, enflamed by the colleague who felt I had made them look 'like muppets', to scour the findings chapters of my thesis for 'their bits'. Reassured that those who had read my thesis understood my overall favourable impression of the team, I had not thought carefully about how individuals might feel when reading my account of their words and deeds, and the inferences I had drawn from them.

When an ethnographer writes up their study, each time they draw on a fieldnote, a transformation occurs – the actions of individuals in the heat of the moment become ossified into a record of the manifestation of some more generalised typology (Atkinson 2015). This occurrence in my work must have been troubling for my participants. For example, social work, especially in countries dominated by neoliberalism in social policy, such as the UK, has experienced a somewhat fractious relationship between theory, as taught on qualifying courses in academic institutions, and real-world practice (Narey 2014). Anti-oppressive practice, which has become a unifying theoretical approach for social workers (Millar 2008), encourages social workers to look beyond the immediate context of an individual's situation and to understand the issues they face as resulting from wider structures and social forces (Dominelli 2002). Social workers are taught that they should seek to counter discrimination and oppression, rather than simply help individuals to adjust to the injustices they encounter. However, such ambitions are difficult to achieve within welfare systems that aim to provide only residual support to the most needy, in which social workers are often obliged to perform the function of being gatekeepers for services (Dustin 2007). In my fieldwork, I had noticed how social workers maintained a strong interest in human rights, social justice and empowerment, yet tended to focus their efforts only at the individual level, rather than agitating for wider change in the manner suggested by anti-discriminatory practice.

When writing up, I gave the example of a social worker who found that a patient's walking frame had been lost when the hospital moved him from one ward to another. I described how the social worker expressed outrage on the patient's behalf and resolved to track the walking frame down and restore it to its owner but did not raise the possibility of raising a complaint with the hospital management that such an oversight had occurred (as it frequently did). From this, I concluded that although social workers were able to stand up for the rights of individual patients, they were not able to engage in attempts to tackle the wider structures that caused disadvantage to individuals.

For the individual social worker who read my account of his work, the ossification of a moment in his daily work to make a general comment on this point may have been experienced as symbolic violence (Bourdieu 1991). As a profession, social work is much maligned, and popular media often convey the message that social workers are not good enough (Butler and Drakeford 2005; Munro 2011). Social workers often fit Power's (2008) description of the 'distressed professional', subject to unsympathetic management systems and held up against unrealistic expectations, and here my work appeared to be telling them, again, that what they do is not good enough. Further, as a male researcher, I was investigating work done predominantly by women, in a profession that was created and developed predominantly by women (Wilson et al. 2011). Commentary on their work by a male academic that they perceived as unsympathetic might be reasonably interpreted as yet another manifestation of male dominance and female subjection. Seen in this light, it is easy to understand how participants would have experienced my study as exploitative.

Ultimately, the charge of exploitation is one I reject. Though I have noted that social work might be considered a distressed profession, it is important to remember that social workers retain a great deal of power over individuals (Dominelli 2002), and it is therefore proper that social work practice should be scrutinised through research. Within the wider context of my study, the observation regarding the absence of attention to wider structures by social workers was sympathetically understood. The picture I attempted to convey was that hospital social workers are subject to intense pressure from managerial systems to accomplish patients' discharges as quickly as possible, and are aware that older people risk suffering nosocomial infection, loss of mobility and loss of confidence the longer they stay in hospital (Hirsch et al. 1990). There is simply too much individual need to be dealt with for social workers in this setting to be able to devote time to any activities that do not have an immediate benefit to an individual patient who is awaiting discharge from hospital. The conclusion I drew from my study (see Burrows 2020) is that the academic expectation of what social workers can do is not always realistic in the settings in which social work occurs, and that UK social work practice is constrained by legislation and managerial systems over which social workers have no control. In fact, I argued that it is encouraging that social workers still find ways to advocate for human rights, social justice and empowerment in such a bureaucratised and high-pressure practice setting as a hospital discharge team. These messages became lost, however, when the participants experienced the discomfort of seeing their words and deeds displayed for permanent public scrutiny.

I do not wish to suggest that my participants struggled to contextualise my records of their actions with my analysis of their work because of a fault in their approach to reading it. I believe the issue arose because I had not reacted effectively to the profound alteration in the relations between me and them when I transitioned from fieldworker to author. In seeking feedback on my work from participants, I was re-entering the field; yet, crucially, the self I was presenting was different. The self I had occupied in the field was a self of the field, shaped by the field. Away from the field, the process of analysing fieldnotes as 'data' creates another self: the author self. Whereas my field self was characterised by a conscious incompetence and deliberate, but authentic, amiability towards others, my author self was a detached thinker, who claimed the privilege of interpreting dispassionately the actions and incidents he had witnessed. It is an unsettling aspect of doing ethnography that we are thus confronted with contradictory versions of ourselves. Presenting as the acceptable incompetent in the field, while systematically analysing the words and deeds of people whom you actively cultivate as experts from whom you can learn, can feel like an act of dishonesty or inauthenticity (see Stacey 1988 or McLuhan 2020). Yet these selves are not creations over which we have full control – they are created partly by decisions we make, yet also partly by the setting in which we find ourselves. The field self is no less authentic because there is an author self, yet returning to the field brings the contradiction to the foreground for both researcher and participants. Wolf (1991), for example, notes how the growing consciousness of his role as an ethnographer detracted from his role as a biker when studying the Rebels motorcycle gang, and led gradually to his exclusion. This highlights the difficulty for the ethnographer of managing field relations once the author self enters the field.

The confrontation of field self and author self was deeply unsettling to me, because it caused me to question whether the self I had originally presented in the field was authentic. In order for it to be possible for me to publish my work, I needed the team to accept my author self. The renegotiation of field relations in this respect occurred through a series of e-mails between me and the gatekeeper, who spoke on behalf of all the participants. The priority for me was to identify exactly where the problems lay for individual participants – which data extracts were a problem, and why? The general issue appeared to be that participants worried about individual actions or comments being taken out of context, leading to what they perceived as an unbalanced portrayal of how they worked. As an author, I promised to examine every issue raised, but also defended my work by pointing out that balance was often achieved over the course of several paragraphs, rather than through intricate discussion of each data extract.

Waiting for the list of objections to return was an uncomfortable time. When it arrived, I was pleasantly surprised at its brevity and, as far as their implications for my arguments and inferences in the work as a whole were concerned, their superficiality (although I acknowledge the points were anything but superficial for the individuals described). There were some factual corrections to be made, which did have some bearing on my interpretations in a few places, and one expansion of an episode of which I had only partial knowledge, which helped me to expand on an interpretive point more thoroughly. I was happy to address each point raised – there was not a single fieldnote about which I felt the need to argue against amendment. I offered to meet with the individuals to talk through the changes I would make, but they were satisfied that I could talk through them all with just the gatekeeper. The meeting between us for this purpose was fairly brief, but cordial. After I had completed the amendments, I sent the chapters through to the gatekeeper to check that they were happy. This marked my final exit from the field.

It is interesting that the participants did not insist on meeting with me face to face so as to talk through the changes I would make at their request. I admit that I was relieved when the gatekeeper confirmed that only she would be present, and suspect that meeting again would have felt as difficult for them as it did for me. As Coffey (1999) observed, conducting and participating in ethnographic research frames the relationships we form with people during fieldwork. Once my 'field self' had been replaced by my 'author self', a continuing relationship between me and the participants was untenable. My desire to seek their feedback was motivated by a wish to collaborate, and to ensure that my work was as accurate and well informed as possible; yet I caused us all unnecessary discomfort because I had not thought through the implications of returning to the field inhabiting a different self. One telling comment in our last meeting was the gatekeeper's reproach of my description of how I had managed field relations, where I had written in my thesis: 'I employed a number of tactics to mitigate the impact of my presence on [participants].' This was, indeed, a poor choice of words, with implications of duplicity that did not in any way reflect my lived reality of managing field relations, and which was a lazy way of signalling to examiners that I had thought 'as a social scientist' about field relations. For participants reading my description of my interactions with them, this must have been a jarring introduction to my 'author self'.

Conclusions

My experience of leaving, and then returning to, the field has provided insight into the way that social life can create conflicting selves that exist authentically depending on the social context. The self is a dynamic, performative process, not a state of being, and its forms coalesce according to people, place and time. Most of the time, we shift between selves smoothly and without giving our fragmentation thought. Doing ethnography can force us to confront the dissonance of different selves that are equally real and authentic, because by its very nature it requires that we encounter the field self from the perspective of the authorial self, and this is profoundly unsettling. The discomfort caused to me and my participants arose because I mistakenly assumed that I had left the field when I finished data collection and did not give enough consideration to the self I was presenting each time I revisited the field, nor to how sharing my written work would transform field relations.

Authentic engagement in respondent validation is not a simple case of politeness or fulfilling obligations to our participants – it is a part of fieldwork that must be crafted and reflected upon as much as any other. It may have a significant impact upon both the researcher and the participants' understanding and memories of field relations. The ethnographer, as author, must consider the impact of their words on participants, and work to manage the shift in field relations that may occur.

Notes

1 Social work with older people, and especially hospital social work in the UK, is paid far less attention in public discussions about social work than child protection work, which is often subject to controversy and public attention.
2 In UK slang, the term 'muppet', a reference to the puppet show *The Muppets*, means an idiot.
3 For a fuller discussion of the ethics of ethnography, see Bosk (2008).

References

Atkinson, P. (2015). *For Ethnography*. London: Sage.
Bosk, C.K. (2008). *What Would You Do? Juggling Bioethics and Ethnography*. Chicago: Chicago University Press.
Bourdieu, P. (1991). *Language and Symbolic Power*. Cambridge, MA: Harvard University Press.

Burrows, D. (2020). *Critical Hospital Social Work Practice*. Abingdon: Routledge.
Butler, I. and Drakeford, M. (2005). *Scandal, Social Policy and Social Welfare*. (2nd edn). Basingstoke: Palgrave Macmillan.
Coffey, A. (1999). *The Ethnographic Self*. London: Sage.
Delamont, S. and Atkinson, P. (2021). *Ethnographic Engagements*. Abingdon: Routledge.
Dominelli, L. (2002). *Anti-oppressive Social Work Theory and Practice*. Basingstoke: Palgrave Macmillan.
Dustin, D. (2007). *The McDonaldization of Social Work*. Farnham: Ashgate.
Fine, G.A. (1980). 'Cracking diamonds: Observer role in little league baseball settings and the acquisition of social competence', in W.B. Shaffir, R.A. Stebbins and A. Turowetz (eds), *Fieldwork Experience*. pp. 117–132. New York: St. Martin's.
Gobo, G. and Marciniak, L.T. (2016). 'What is ethnography?' in D. Silverman (ed.), *Qualitative Research*. (4th edn). pp. 103–119. London: Sage.
Goffman, I. (1974). *Frame Analysis: An Essay on the Organization of Experience*. New York: Harper and Row.
Guba, E.G. and Lincoln, Y.S. (1994). 'Competing paradigms in qualitative research', in N.K. Denzin and Y.S. Lincoln (eds), *Handbook of Qualitative Research*. Thousand Oaks, CA: Sage.
Hammersley, M. and Atkinson, P.A. (2019). *Ethnography: Principles in Practice* (4th edn). London: Routledge.
Hickey, A. and Smith, C. (2020). 'Working the aporia: Ethnography, embodiment and the ethnographic self'. *Qualitative Research*. 20(6): 819–836.
Hirsch, C.H., Sommers, L., Olsen, A., Mullen, L. and Winograd, C.H. (1990). 'The natural history of functional morbidity in hospitalized older patients'. *Journal of the American Geriatric Society*. 38(12): 1296–1303.
McLuhan, A. (2020). 'Feigning incompetence in the field'. *Qualitative Sociology Review*. 16(1): 62–74.
Mead, G.H. (2011 [1913]). *A Mead Reader*, ed. by F.C. da Silva. Abingdon: Sage.
Millar, M. (2008). '"Anti-oppressiveness": Critical comments on a discourse and its context'. *Journal of Social Work*. 38(2): 362–75.
Munro, E. (2011). *The Munro Review of Child Protection: Final Report, a Child-centred System*. London: The Stationery Office.
Narey, M. (2014). *Making the Education of Social Workers Consistently Effective: Report of Sir Martin Narey's Independent Review of the Education of Children's Social Workers*. London: Department for Education.
Plankey-Videla, N. (2012). 'Informed consent as process: Problematizing informed consent in organizational ethnographies'. *Qualitative Sociology*. 35(1): 1–21.
Power, S.A.R. (2008). 'The imaginative professional', in B. Cunningham (ed.), *Exploring Professionalism*. pp. 144–160. Bedford Way Papers. London: Institute of Education Publications.
Roth, J.A. (1962). 'Comments on "secret observation"'. *Social Problems*. 9(3): 283–284.
Stacey, J. (1988). 'Can there be a feminist ethnography?' *Women's International Studies Forum*. 11: 21–27.

Wilson, K., Ruch, G., Lymbery, M. and Cooper, A. (2011). *Social Work: An Introduction to Contemporary Practice*. (2nd edn). Harlow: Pearson Longman.

Wolf, D.R. (1991). 'High-risk methodology: Reflection on leaving an outlaw society', in W.B. Shaffir and R.A. Stebbins (eds), *Experiencing Fieldwork: An Inside View of Qualitative Research*. pp. 211–223. London: Sage.

16

Commenting on legal practice: research relationships and the impact of criticism

Daniel Newman

Introduction

If *How to Lose Friends and Alienate People* was not a book title already taken by one of the most morally repugnant minor members of the UK's commentariat, I might have made it the title of this chapter, because that was what happened to me. Here, I am talking about how my ethnographic research introduced me to a community that I embraced and that felt like I became a part of over the course of a year. In reality, I was just engaging in research and, unfortunately, that research did not end up depicting the people I was spending time with in a good light. This chapter pertains to my first book (Newman 2013). It explores the way I accepted the need to criticise those I was studying, and the consequences I faced.

A central issue considered in this chapter is how I dealt with the situation in which I felt compelled to comment critically on the work of a group I had been following for research and with whom I had grown friendly. The chapter will look at the difficulties of balancing being honest to the research with the pressures that come from interpersonal relationships, which were especially prominent in this study that became known for taking a negative line about the practices of those being studied. The chapter also looks at the fallout in terms of research anxieties caused by critical commentary on research among the communities being studied and subsequent difficulties in recruiting from these communities for future studies.

The research

Decades of underfunding, exacerbated by the UK government's austerity programme from 2010, have gutted the criminal justice system of England and Wales, posing great threat to access to justice. Legal aid was cut by 8.75 per cent under the 2010–15 coalition government (having already been effectively cut under New Labour by 12 per cent); the overall Ministry of Justice budget fell by 29 per cent, the largest cut to any Whitehall department. Such cuts undermine access to justice, which broadly relies upon the availability of criminal defence lawyers funded by legal aid to help suspects and defendants understand a confusing, hostile system where the state inherently has the upper hand. In this context of cuts, I considered it of great importance to investigate the practice of criminal legal aid lawyers, especially so as to understand the service that their clients – defendants relying on publicly funded defence when challenged by the state – receive in austere times.

For this research, I was given the chance to spend a year doing fieldwork looking at access to justice. I wanted to understand how those people suspected and accused of crimes experienced the criminal justice system. Carlen (1976) has shown how the criminal courts function as a form of social control to render defendants as 'dummy players', while McBarnet (1981) documented the way that criminal courts 'alienate' defendants. The defendant can be faced with a bewildering environment once drawn into the criminal justice system. Defendants can quickly find themselves very much out of their comfort zone – left inarticulate and overwhelmed. As such, it seems necessary to provide them with a helping hand to guide them through: an expert, a defence lawyer.

My position here was guided by Felstiner's (2001: 191) contention that 'the production of justice might be defined as a dimension of the relationship of lawyers to clients'. This vision of access to justice depends on the health of the lawyer–client relationship. Access to a legally aided lawyer is not enough of itself. Only if the lawyer appears committed to their client will they facilitate access to justice. To judge this commitment, I wanted to get to know the lawyers as best I could so that I could start to see their realities from within the world of legal practice.

I thus set myself the aim of drawing out how the lawyer–client relationship worked in practice. In designing my research, I was influenced by Goffman (1990) and his dramaturgical metaphor, with the actor putting on a show for the audience. I wanted to get access to these performances. As such, so as to understand the lived reality of practice, the everyday interactions that constituted the lawyer–client relationship, I decided to use ethnography. I was motivated by the belief that this approach represents one of the most valuable contributions that sociology can make to a socio-legal understanding of the lawyer–client relationship. From the start, I wanted to produce a book that would show the reality of that relationship. My hope was to paint a picture that those involved would recognise, so I needed to find a way to integrate myself into a community of criminal legal aid lawyers.

Ethnography had been used before in looking at the lawyer–client relationship in criminal justice and, indeed, was the backbone of the two foremost studies that I was looking to follow on from. These works were *Standing Accused* from McConville et al. (1994) and *The Reality of Law* by Travers (1997b). McConville et al. (1994) used ethnography to report activity from within forty-eight firms for a total of 198 weeks, spending no more than two months at any one firm. Travers (1997b), in contrast, deployed ethnography to address one firm over a period of four months. One of the most notable aspects of these studies is the way they offer conflicting and quite divergent results – and this is where my research picked up.

A condemnation on the existent provision of defence lawyering was provided by McConville et al. (1994), with firms generally neither committed to clients nor organised in a manner conducive to serving them. Lawyers were shown to hold low opinions of their clients and believe them to be undeserving of good treatment. The picture emerged of a deficient profession offering inadequate representation. In contrast, Travers (1997b) presented a contradictory situation; a reading in which defence lawyers are given the opportunity to set their own narrative and, in so doing, present themselves in a far more positive light. By this, highly knowledgeable and competent lawyers were seen to dedicate themselves to their clients. This glowing account found deeply satisfied clients receiving a high standard of care, and – in a thinly veiled dig at McConville et al. (1994) – concluded that researchers need to give lawyers more credit and respect.

It is the work of McConville et al. (1994) that has been most influential, becoming something of the received view of this branch of the legal profession in socio-legal circles. Not only did this study have a greater scope than Travers (1997b), it also chimed with other critical work of the time such as Mulcahy (1994). Meanwhile, Travers (1997b) has not found a particularly wide or receptive audience. This is even though Travers (1997a) presented a sustained attack on McConville et al. (1994), initiating a heated exchange with (the same authors reconstituted as) Bridges et al. (1997) in the *British Journal of Criminology*. In this dispute, Travers (1997a) took issue with the methodological approach of *Standing Accused*. He claimed the work to be informed by the structuralist credo, supposedly guided by 'the correspondence theory of truth', which sees reality existing independent of the observer, providing researchers the opportunity to capture and disseminate it. In stark contrast, he was said to subscribe to an interpretivist epistemology, informed by 'the congruence theory of truth', positing the existence of multiple realities, meaning that the researcher has no privileged position and that social knowledge represents the shared understandings of a community of individuals.

While each of the ideological positions taken by these studies has its strengths, I was concerned that too close an adherence to either might work to limit the scope of my research. I wanted to somehow reconcile them. I followed Hekman (1983: 193) and her plea for socio-legal research which can 'bridge the gap' between structuralism and interpretivism. For Lacey (1994), socio-legal research should have an 'integrated approach', and that is how my project was constructed. I conducted a year of fieldwork with four months spent at three different firms – spending as much time as Travers (1997b) spent at one firm at multiple firms, allowing for some manner of depth and generalisation. So as not to fall into the trap of talking for lawyers, the participant observation in this research was followed by a series of formal interviews with the same lawyers. This provided an opportunity for them to present their own side of the story. In reporting these data, observation and interviews were afforded their own delineated sections, with minimal critical engagement, organised to provide representative accounts of both what was seen and what was said.

I designed my research to reconcile the discrepancy between the damning indictment of McConville et al. (1994) and the cheerful buoyancy of the lawyers in Travers (1997b), supposing that it would reveal that some lawyers were positive about clients, others were not, resulting in a mixed service provision. However, this was not the picture which emerged, and this research combined the two previous studies in a quite different way. In interview, lawyers revealed a client-centred practice, presenting a very healthy image of the lawyer–client relationship. This chimed with that of the lawyers who discuss their practice in Travers (1997b). Under observation, however, there emerged a distinctly lawyer-centred practice, raising great concerns for the health of the lawyer–client relationship. The results were akin to those of McConville et al. (1994). It seemed that, in this situation, lawyers could talk the talk but not walk the walk. It was in my need to document this finding that the problems of leaving the field had their roots. I had spent a long time trying to become one of these lawyers in order to understand their worldviews, such that, once I needed to step back again and judge them, I encountered a problem.

In the friend zone

Over the course of my research I felt that I became close to many of the lawyers I was studying. Tillmann-Healy (2003) has outlined the 'friendship as method' approach and how it allows researchers to know those they are studying in meaningful ways. For Owton and Allen-Collinson (2013), friendship in research is more accurately titled a methodology, with a philosophical underpinning based on challenging the power imbalance between researcher and participant. The use of friendship in research introduces an ethic of caring that reduces the researcher–participant hierarchy and evens out the power balance of that relationship towards something more mutual. This insight came to me only when looking back on my findings. It was not an approach that I had considered when initially embarking on my research. I had not considered its appropriateness when I was studying an ostensibly powerful group in lawyers and was explicitly doing so in order to tackle power imbalances which I found in the lawyer–client relationship that they were already engaged in.

Nevertheless, as is inevitable in ethnographic research where the researcher spends a considerable period in the field, friendships developed. Yow (2018) has argued that research can have positive emotional impacts for those being interviewed, and Mitchell (2019) suggests that sharing painful stories with an 'enlightened witness' in interview can help to make those interviewed feel better when dealing with painful situations. Thus, a feeling of camaraderie developing between researcher and researched is understandable. Ellis (2007) identifies that ongoing and overlapping relationships in research may make loyalties, confidences and awareness of contexts much more difficult to negotiate. Even short of adopting friendship as method or methodology, taking an approach to research grounded in fostering good relations should mean being aware of the value, significance and implications of approaching participants and interactions with participants from a 'stance of friendship' (Tillmann-Healy 2003: 745). This entails what should be principles that underpin any ethical qualitative research project, but that merit recognition, and placement at the forefront, during the often complex and confusing relationships that develop during research. I think here of values such as the researcher treating participants with respect and dignity, fostering empathy and sensitivity and giving voice to participants to treat their stories seriously.

Consideration around friendship came into my project from the way I was so immersed in the community I was studying that lines between research and personal lives became blurred. It was often difficult to decide when I should be in 'researcher mode' or not. Occasionally, lawyers would tell me something was 'off the record' and provide me with some gossip or acerbic opinions. More often than not, they appeared to forget that I was conducting research, and I had to negotiate a moral tightrope as they regaled me with some story that I would assume they did not want repeating but that I had not been explicitly given instructions to withhold from the write-up.

The balancing act was difficult enough when interactions took place in what were more obviously 'work' spaces such as courtrooms, interview rooms and meeting rooms. The ambiguity of spaces that were in or adjacent to their places of work made this division harder again – areas such as advocate's rooms, cafeterias and car trips to or from meetings. I drew the line at social interactions that I was invited to. At the first firm, there were solicitors with whom I would often go out for drinks; at the second firm I would go to barbeques and dinner parties at the houses of the partners, while the third firm invited me to their Christmas party at a local hotel. It was difficult to not report any events from the latter occasions, especially when, with their guard down, I would get the juiciest titbits. For example, my book showed the poor regard in which some of these lawyers held their clients, based on things said while going about their jobs; if I had let myself use things spoken when they were off duty, the point about how lawyers saw their clients would have been far stronger and would probably have led to a more exciting book. I had more than enough data – especially those salacious, critical quotes that would be so attractive to readers – from the official encounters though, and therein lay the central problem.

The impact of these friendships was that, when I got to the end of my fieldwork and started to look back over my data, I was conflicted. I had liked them, I enjoyed being with them, but there was a problem that I had been trying to avoid facing up to while we were knocking about together day in, day out. In analysis, one thing leaped out at me again and again: these lawyers were treating their clients terribly. I did not want to face up to the consequences of this narrative; these were people who had welcomed me into their world, whom I found funny, whose company I had relished, who were kind to me, whom I had become invested in over the course of the years. For O'Reilly (2009), one danger for ethnographers is to become too involved in the community under study, thus losing objectivity and distance. I was worried that this was where I found myself.

The problem

When writing up this work I had to try to reconcile these two contradictory accounts of the lawyers' attitudes towards their clients. Lawyers claimed in interview that relations with clients were positive, even though the participant observation suggested that they were negative. I set myself the task of answering the question: 'how can lawyers claim to hold positive attitudes yet actually display negative attitudes?'

One approach – which might seem obvious to many outside the study – is that a party has lied. In a case of such divergent findings, this would offer an immediate means of understanding the schism. First and foremost, it could be suggested that I was lying. The lawyers' accounts in the formal interviews represented the truth of the situation; they had positive attitudes. What I had disseminated from the observation was either, at best, distorted, or, at worst, fraudulent. By this line, there was a healthy lawyer–client relationship; it was all sweetness and light. I had twisted things for my own ends, perhaps bent on some lawyer-bashing crusade; the idealised – negative – representation of lawyers which Travers (1997a: 370) had criticised McConville et al. (1994) for constructing.

Those were charges I could have easily defended myself against. I might offer my notebooks – the dozens of volumes recording the events that I had witnessed. However, that would prove little more than that I had at some point handwritten the words that appeared in the book. All I could contribute were sentiments, the foremost of which was the assurance that I had sought to abide by the ethical code of the university law school from which the research was undertaken, as well as working within my own moral framework to be as honest as possible. In so doing, I had worked tirelessly at collecting the data for this research and, frankly, I didn't see the point of putting in all that effort when I could have just stayed at home with a nice cup of tea and made it up in a fraction of the time. Ultimately, though, any external evaluation was out of my hands.

The others who might be accused of lying were the lawyers themselves. The negative attitudes that I had reported from the participant observation were accurate; however, the lawyers' response in the formal interviews was an attempt at deception, subterfuge. This would run the common-sense line that friends and colleagues typically comment on hearing of my data: 'what else would they say?' It would be inevitable that lawyers would seek to justify what they did and protect themselves from criticism – the lawyers knew they had not met the standards demanded of them. These expectations can be gauged from the Code of Conduct for Solicitors, which attributes their central responsibility to clients. Considering that lawyers are primed with such duties, it is likely they would be alert to the need to deflect any claim that they had neglected their commitments.

There may be some validity in this position, and the research can conceivably be read in this manner. Lawyers were off guard for much of the extended period of participant observation. They accepted me and largely acted naturally. The lawyers had little control over how or when – or even whether – what they said or did was documented in my fieldnotes. They did not know what I was recording about them, but the formal interviews offered a chance for them to paint themselves in a positive light. These were audio recorded, meaning that lawyers were captured for posterity speaking on tape and, thus, could be played back to anyone who cared to listen; all that they said there clearly counted. In this artificial setting, they bore witness to this evidence collection, and so were highly attuned to my role and what they should say. In these circumstances, they fed me lines to portray themselves as they wanted and so took control of managing the image I saw. They would expressly avoid disseminating negative attitudes towards their clients, aware of the bad light this would cast on them professionally. Accordingly, it was plain that I should be told something different from what I saw; lawyers would not knowingly contribute to their own mortification.

I think that most readers of my book from outside the legal profession supposed the latter, that the lawyers were lying. In subsequent years, I had academics, campaigners and journalists come up to me and say how brave I had been for showing the truth of how lawyers treated their clients. They all read my account as though I had clearly presented lawyers as lying. But I did not. My book had actually gone on a convoluted alternative explanation of several thousand words to try to avoid pinning the blame on lawyers. I noted that it would make me feel personally uncomfortable to accept that lawyers were lying, and acknowledged that I might have lost impartiality, become sympathetic with those I had been studying and reached a point where I declined any negative analysis of their lives as per O'Reilly (2009) – the process sometimes (albeit problematically) referred to as 'going native'. But I also suggested that, on an academic level, the 'lying lawyer' trope was something of a hasty, knee-jerk attempt at an explanation anyway.

As such, I explored a proposition that I said could be read either alongside or in place of the suggestion that lawyers were deliberately lying. This discussion entertained the possibility that lawyers were able to profess positive attitudes while concomitantly displaying negative attitudes. They could do this without necessarily being cognisant of the contradiction. This notion rests upon the lawyers learning to distinguish between *the* client – in general terms – and *a* client – as particular example. In so doing, I analysed attitudes in psychological terms, positing that lawyers learned to adopt specific attitudes towards their clients. This was followed by a discussion of how the lawyers might have come to develop these attitudes, focusing on their self-image and the manner in which this was challenged by their position within the legal profession. Ultimately, I used this to suggest that positive and negative attitudes could coexist, but that this jeopardised the health of the lawyer–client relationship.

Drawing on Freud and Jung was an interesting academic exercise, and I went on to develop it further in subsequent papers that were fun and – as far as I am concerned – produced some valuable insight for lawyer–client relations. I am still unsure, though, just how much of this was an attempt to avoid admitting to myself that the lawyers may have lied to me and that these people whom I found so attractive were treating clients, those who relied so totally on their lawyers for justice, so badly. I was doing this research for those suspected and accused of crimes – these were the people I care about – but I seemed to have been working awfully hard to give a pass to the lawyers whom I had seen act reprehensibly to them. I seemed to feel some sort of obligation to these lawyers based on our friendly relations.

Leaving the field

Mungham and Thomas (1979) in their own research with criminal lawyers have discussed the problems of studying the 'locally powerful'. The danger of aggravating them with the results is massive. Luckily, I moved away from my research site shortly after publication. But, despite my mealy-mouthed attempt at prevaricating over what I had found, I am aware that many lawyers were very angry on reading my book: leaving the field led me to a world of criticism of how I had treated lawyers from within the legal profession.[1] I have been shown social media discussions of my work that offered furious rebuttals of my analysis from lawyers questioning my competence and integrity. I remember being hurt by the snobbishness of some of these comments, especially when I published my book during my first academic appointment as a research assistant. It was made clear to me that some considered that my apparently lowly professional status gave me no right to comment. They did not think that what I was showing was fair or accurate, and blamed my anti-lawyer prejudice. I found this shocking, as I had gone into the research looking to show how important lawyers were – and I still felt like I was on their side, and that the main target of my attacks, indeed, the only group I explicitly was gunning for, were the governments who had denigrated and debased legal aid over the previous two decades through cuts and media campaigns.

I was told in no uncertain terms that it would be very difficult for me to do work on lawyers again because people would be reluctant to take part, in case I performed some kind of 'hack job' on them. I was worried that what was going to be the main area of my research was suddenly cut off for me. Some lawyers who knew my work would not recommend colleagues to do research with me in case I upset them. During one recent interview with a lawyer, they attempted a 'gotcha' moment where they brought up my past research that they had been shown and launched into an invective on why I was unfair. Lawyers that I know from other areas were generally a little reserved when talking about the book, and I often had the feeling that they were unsure of me after what I had done. I felt sad that people might be making negative judgements of my character because of this work. I was terrified to ever come across any of the lawyers whom I worked with in case they had read the book and felt misrepresented, or were generally wanting to vent frustrations at me. I hate being talked about behind my back, and so could imagine how hurt I would have been reading accounts about myself that I did not agree with or have any control over – and so I could imagine how they would feel. I have pulled out of events and projects that would have been beneficial to my career, in order to avoid such awkwardness.

The impact endured into my subsequent research. I carefully spent several years avoiding doing more research with lawyers so that I would not be put into this situation again. I finally realised that there was important research to be done on defence lawyers that nobody else was doing. And for most people who know my research, they considered that I should be the one who filled that gap. I thus had to give much thought to how I designed it. I had left the city where my previous research was done, and so had new groups of lawyers whom I could work with (though even now, some lawyer friends are reluctant to give me introductions because of my reputation). Eventually, I conducted a series of interview-based studies, despite my conviction that interviews can be limited as a social research method and that, for the best results, they should be complemented by some manner of observation. I wanted to do observational work, but I always stopped myself when recalling my previous experience.

This has been the biggest impact of leaving the field, the way it has restricted me in my ongoing research. I am proud of the work I have done on access to justice since that first book; it all adds to my intention of promoting understanding on how access to justice is experienced. However, I am acutely aware that, as a legacy from my first book, I am reluctant to do the kind of hard-hitting exposé of criminal legal aid that could be done. I never intended to be uncomplimentary to the lawyers themselves and I still believe that I was more critical of how systematic, structural factors had influenced their behaviour. Even if they were lying, and even if they did treat clients badly, they were (to some degree) pushed into this by legal aid cuts and the immense pressure of working in this little-loved part of the welfare state with a clientele whom society has largely abandoned and decided are unworthy of good treatment. I now feel that I must be careful to present this just right, so that the emphasis is so obviously on the government, austerity or neoliberalism that my message does not end up getting lost under a whole of bunch of upset lawyers piling in because I have hurt their feelings. I am compelled to do this almost to the extent that I might deny agency to the lawyers, lest people take the wrong message from my work and think I am purely intent on a lawyer-bashing vendetta. It involves an awful lot of over-thinking my analysis.

Such navel-gazing might have been avoided if I had had a clearer sense of the research journey from the outset. I was a novice researcher, and likely lost track of the end whereby my data would need to be released into the world. I may have been naive, excitedly enjoying engaging in the fieldwork for its own sake, when a more 'switched on' researcher would have had sight of where they were going. Maintaining an awareness that the experiences would be written up and told to others might have helped to preserve boundaries, meaning that I did not find the transition so difficult. Others new to ethnography will likely find themselves in similar situations. I recommend that they give thought to this issue of how to present findings honestly, from the start of their project. New ethnographers need be aware that things might get awkward and may be uncomfortable when they come to write up research, so they need to manage that in terms of the relationships they develop and their own self-understanding.

From the outset of an ethnography, I would suggest that thought should be given to whether and how a researcher will have the confidence to offer their own reading of a situation without fear of upsetting those whom they studied. It is plausible that spending a long time with groups can lead to friendship – or at least some sense of respect or feeling of obligation – and that can be hard when you subsequently need to criticise them. I have no set of rules to follow, but want to encourage others to have the foresight that I lacked, to help them and their research.

Note

1 Other ethnographic studies with lawyers such as Pierce (1996) have also faced critical reactions from practitioners.

References

Bridges, L., Hodgson, J., McConville, M. and Pavlovic, A. (1997). 'Can critical research influence policy?' *British Journal of Criminology*. 37(3): 378–382.

Carlen, P. (1976). *Magistrates' Justice* (London: Martin Robertson and Co).

Ellis, C. (2007). 'Telling secrets, revealing lives: Relational ethics in research with intimate others'. *Qualitative Inquiry*. 13(1): 3–29.

Felstiner, W.L.F. (2001). 'Synthesising socio-legal research: Lawyer–client relations as an example'. *International Journal of the Legal Profession*. 8(3): 191–201.

Goffman, E. (1990). *The Presentation of Self in Everyday Life* (London: Penguin).

Hekman, S. (1983). *Weber, the Ideal Type and Contemporary Sociological Theory*. Oxford: Martin Robertson and Company.

Lacey, N. (1994). 'Introduction: Making sense of criminal justice', in N. Lacey (ed.), *A Reader on Criminal Justice*. pp. 28–34. Oxford: Oxford University Press.

McBarnet, D.J. (1981). *Conviction*. London: Macmillan.

McConville, M., Hodgson, J., Bridges, L. and Pavlovic, A. (1994). *Standing Accused*. Oxford: Oxford University Press.

Mitchell, D. (2019). 'Oral histories and enlightened witnessing', in K. Moruzi, N. Musgrove and C. Pascoe Leahy (eds), *Children's Voices from the Past: New Historical and Interdisciplinary Perspectives*. pp. 211–231. London: Palgrave Macmillan.

Mulcahy, A. (1994). 'The justifications of justice'. *British Journal of Criminology.* 34: 411–430.

Mungham, G. and Thomas, P. (1979). 'Advocacy and the solicitor-advocate in magistrates' courts in England and Wales'. *International Journal of the Sociology of Law.* 7(2): 169–195.

Newman, D. (2013). *Legal Aid Lawyers and the Quest for Justice.* Oxford: Hart.

Owton, H. and Allen-Collinson, J. (2013). 'Close but not too close: Friendship as method(ology) in ethnographic research encounters'. *Contemporary Journal of Ethnography.* 43(3): 283–305.

O'Reilly, K. (2009). *Key Concepts in Ethnography.* California: SAGE.

Pierce, J. (1996). *Gender Trials: Emotional Lives in Contemporary Law Firms.* Berkeley: University of California Press.

Tillmann-Healy, L.M. (2003). 'Friendship as method'. *Qualitative Inquiry.* 9(5): 729–749.

Travers, M. (1997a). 'Preaching to the converted? Improving the persuasiveness of criminal justice research'. *British Journal of Criminology.* 37(3): 359–377.

Travers, M. (1997b). *The Reality of Law: Work and Talk in a Firm of Criminal Lawyers.* Aldershot: Ashgate.

Yow, V. (2018). 'What can oral historians learn from psychotherapists?' *Oral History.* 46(1): 33–41.

17

Emotional honesty and reflections on problematic positionalities when conducting research in another country

Ashley Rogers

Introduction

This chapter contributes to research and ethical discussions around leaving the field and the shift from the fieldwork stage to writing up. While discussions on ethics tend to be focused on the 'pre-fieldwork' and 'in-the-field' stages, this chapter explores personal and professional tensions experienced post-fieldwork, in the liminal space between 'the field' and 'home'. This chapter is a balancing act of self-analysis and reflection alongside practical and academic discussions on methods, ethics, positionality and writing, but, importantly, it is also an attempt to emphasise the complexities of the emotions that are at play in qualitative research and the need to centre these. This book has provided space to address what I have always left unsaid in a written form, aside from in my personal fieldwork diary, from which I include some excerpts similar to what Humphreys (2005) refers to as 'narrative vignettes'. Like in Humphreys' vignettes of stress during his career, I use the excerpts to transport and connect you to the situated emotions at the time they were experienced. Throughout this chapter it becomes clear that leaving the field occurs not just physically, and mental shifts are also required. Although the physical exit from the field took four plane journeys over two days, the mental separation was far more challenging, never complete, and that legacy lives on.

I begin the following section of this chapter by providing an insight into the context of my research and, in doing so, the tensions I experienced begin to emerge. The next section explores my feelings of unease around my positionality and the effects it had on returning to data and writing, and the final section pinpoints some tools and techniques that helped me to overcome some of the challenges I experienced, with other techniques woven through the fabric of the chapter.

The context

In 2014–15 I conducted ethnographic research exploring men's violence against women and women's (in)access to justice in Bolivia. My interest in Bolivia can be traced back to my first MSc in Human Rights and International Politics (2008–9), where I learned of Bolivia's New Constitution and the involvement of women's groups and Indigenous groups in its development. I continued to follow these developments, learning about the history and politics of the country and, in 2015, an Economic and Social Research Council-funded PhD afforded me the privilege of spending one year in La Paz to conduct an ethnography of Bolivian women's legal consciousness. Legal consciousness, in its simplest form, refers to the way that people engage with the law, including their awareness and understanding of it, and their relationship with and to it in their everyday lives. The research included participant observation in a women's centre, gathering life stories and conducting semi-structured interviews with non-governmental organisations (NGOs) and government officials. While I had some reasonable skills in speaking Spanish, I took lessons throughout my PhD, including while I was in La Paz. Doing so enabled me to make initial connections for my research and build friendships, an important element of adjusting to life in a city, culture and language very different to my own. These relationships with people were also vital for checking my interpretations while in the field, exploring interview questions and reviewing cultural appropriateness, as well as for gaining deeper understandings about Bolivian society. As my sole purpose for being in Bolivia was to conduct research, I was always 'switched on' to possible opportunities for data collection (to Bolivian politics, to people's language use, to all interactions), and while I eventually developed my own way of navigating work and 'time off', the boundaries were always blurred between the positions of 'doing fieldwork' and just 'living in Bolivia'. In both positions I was always an 'outsider', and this impacted far more than I had anticipated on how I felt not only about the research but about myself as a researcher, even after leaving the field.

An outsider

Here, I explore two tensions that were encountered not only *in* the field but upon leaving. As the physical distance between myself and 'the field' grew, so too did my concerns around my privileged position as an outsider from the Global North. Combined with my emotions connected to the stories of violence that I heard, I also avoided my data and writing. Prior to beginning my fieldwork, I produced a lengthy ethics document that

Emotional honesty and problematic positionalities 245

engaged with the practical elements of gaining access, gatekeepers and informed consent and discussed issues around researcher positionality, power dynamics, representation and culture shock. In this sense, it incorporated both 'procedural ethics' and 'ethics in practice', between which there is often a disconnect, particularly in formal institutional review boards (Guillemin and Gillam 2004). The ethics document included a section on 'leaving the field', but I am rather embarrassed to admit that it was the shortest of them all as I gave it very little thought. It seemed so far away. In hindsight, it deserved far more space and consideration.

As I delved deeper into the methodological literature during the first year of my PhD, I developed a deep discomfort around my position as a researcher. I had previously conducted research on politics and women's movements in Bolivia, but that was literature based. I also explored the construction of the legal subject in 1500s Latin America but that, too, was literature based. I was (am) a Global North researcher exploring an issue in the Global South. I was not the first, but did that make it OK? I did not refuse the opportunity to conduct the research project I had outlined and for which I had been granted funding, but I vowed to be aware of my privileged position, to ensure that I was being reflexive throughout, and thus I proceeded more cautiously and nervously than initially planned.

Despite exploring 'positionality' through the literature before I began fieldwork, I did not think about the ways it would be experienced during fieldwork *and* once I left. The move from literature to practice was complex. Positionality in research refers to the different social categories of differentiation that construct and identify a person, and with which a person may identify. Like identities, these positions are not necessarily fixed, and this is especially the case when it comes to immersive methods like ethnography. For many ethnographers there are often challenges in relation to identity and positionality, and a common one is that of 'insider-outsider'. In Bolivia though, I was an outsider, in every way possible.

While my PhD proposal was primed to explore women's legal consciousness in a broader, more theoretical sense, this did not seem doable in the way I had imagined it pre-fieldwork. As I met women's groups and negotiated access shortly after my arrival in Bolivia, I was welcomed, often with great enthusiasm, but it quickly became clear, from long silences, sideward glances and puzzled looks, that my research topic was not being well received. It was too ambiguous. I realised that my proposed project was perhaps only appealing to my own more philosophical socio-legal interests rather than to the particularities of the situation for women in Bolivia and their (in)access to justice. The funding and time constraints of a PhD in the UK meant that I panicked a little about the feasibility and meaning of my work and the possibilities of changing it. One month after my arrival in La Paz, though, on the International Day for the Elimination of Violence against Women, and after deeper engagements with women in the city, I refocused my research to explore Bolivian women's access to justice and the new law to eradicate violence against women, Law 348 (see Rogers 2020). While, theoretically, legal consciousness still played a central role, the focus on something more tangible, Law 348, centred the importance of women's experiences of injustice and provided a clearer topic of discussion.

Combining a more flexible methodological approach to ethnography with the voices and values of the people being studied – rather than feeling entirely constrained by the initial research proposal – is central to feminist research (Hirsch 2002; Daly 1997). Hirsch (2002: 14) points out that 'the techniques of feminist research routinely question and blur the boundaries between observer and observed' and, in doing so, research projects and their methodologies are defined and redefined not only by the researcher but also by those being researched (Burgess 1995). By having open discussions about the focus of the research with women in Bolivia, I attempted to draw on more of a collaborative or participatory ethnographic approach (Lassiter 2005; Huisman 2008). In fact, I felt myself clutching at these techniques and the literature around them, wishing I had adopted more of a participatory action research approach from the beginning.

Indeed, I often questioned whether I should be doing the research at all. But there I was, in La Paz, 'doing' the research. After my slight project alterations, further encounters with participants and potential participants were smoother, easier, with no more puzzled looks. Women were keen to discuss their experiences and have me spend time in their company. But I always felt, and as Griffiths (2017: 7) suggests, that some of these tactics, which are employed to try to balance power dynamics or positions, are often mere forms of 'self-satisfaction in a somewhat too-easy passage from elite institution to periphery'. My feeling of unease regarding being an outsider was always there, and it popped up during one particular exchange:

> I was at a Bolivian friend's house for dinner tonight and her father asked why I would choose to study women's rights and violence in Bolivia (a question I receive at least once a day if meeting someone new), when in fact the problem existed the world over, and indeed in my own country. And, although he did not say it, I believe he also wanted to ask, 'Why you? What right do you have to study this, here?' I don't know if this is my own insecurity about my position or not. I am not sure, but it was certainly what I had deduced from his question because I was sure I could detect at least a hint of hostility behind it. I think I will talk about this tomorrow when I go to the centre [where I was doing participant observation]. (Excerpt from fieldnotes after returning from dinner, during what I had considered to be my non-working 'down time')

After recounting this story to some of the women the next day, they brushed it off because he was a man, and told me that they were surprised I had not already experienced this from other men. In their eyes, I was conducting this research because I was a woman, and it was *that* part of my positionality and identity that, for them, was important (Reyes 2020). So, I carried on. In fact, it was only when I was leaving Bolivia that a lot of my discomfort about my position came to the fore. Soon, I no longer had data collection to occupy my mind, nor such easy access to the women's reassurances. I had not realised quite how vital they were for my ability to push through the tensions I was feeling. As an outsider, I always believed it would be easy to leave the field and return home to complete analysis and writing, but after saying goodbyes during various 'despedidas' (leaving parties), I was consumed by sadness, guilt and anxiety. Sadness at leaving behind people I had come to care for so deeply and a city and culture I had come to love, guilt at being able to leave and anxiety at the prospect of the next phase of research.

When conducting research in another culture, the socially constructed nature of the research experience is more obvious, and the positionality of the researcher is central to this. Sultana (2007: 375) points out that:

> [A] key concern in pursuing fieldwork that has plagued critical/feminist scholars is the issue of representation, where over-concern about positionality and reflexivity appears to have paralyzed some scholars into avoiding fieldwork and engaging more in textual analysis.

This feeling of being 'paralysed' accurately reflects my state of mind when I left Bolivia and returned home. Sitting on the plane, watching people board and waiting for the crew to prepare for take-off, I put my laptop on my knees and tried not to cry. As I opened, closed and moved files to their rightful place, I continued to make notes about some of the broad themes I had identified across the data during my time. What I should have done was give myself time to think, to compose myself and, in fact, to grieve, because ultimately I was leaving people and a place that I had grown incredibly fond of. But, trying to put my 'professional head' on, I opened the life stories and interviews and played short audio clips through my headphones. A wave of nausea washed over me. With the plane ready for take-off, I was relieved to put my laptop back into my bag. I took out my fieldwork diary instead:

> I feel sick. It's not the altitude this time, or the lack of oxygen. Having been here for one year I've become accustomed to that. What is this feeling though?! It's heavy, like breathlessness.
> Like loss,
> like fear,
> like dread.

I'm on my flight back home now, but what is 'home'? I've worked so hard to not only gather data, but to build a life here in Bolivia that would enable me to mentally and physically conduct this research for which I have been funded. I built a life in 'the field', one which I am sad to leave behind. The strong women I leave behind. The injustices. The marches and the movements. So yes, it's sadness. But that's not the feeling, or at least it's only part. What is it?!

It's not just my anxiety, and it's big now, pulsing hard through me. I can still see snow-capped mountains beneath the clouds, the ones that once towered over me. I can see the long stretches of the Altiplano. But it's all shrinking away. My fieldwork is over. I'm really leaving.

I am now face-to-face with the exact stage I have been dreading, going home and writing. I feel sick. Did I do the right thing? Who am I to speak about these women's lives? How will I write this? Should I write this? (Excerpt from fieldnotes as I departed on the plane)

Returning

Ethnographers are often at the margins between two cultures where they decode, interpret and translate the host culture based on their knowledge and understanding of their home culture, which inevitably influences their knowledge and understanding of the host culture. It is a cyclical process, and I was caught both physically and mentally between these two spaces, 'the field' and 'home'. The cause of much of my anxiety was the power I held in representing women's stories, and how I could do them justice (see Tuhiwai Smith 2004). Despite checking interpretations with the women during fieldwork, the position of power to document and (re)present the data lay with me, a position that cannot easily be shifted, regardless of attempts to do so (Stacey 1988; Griffiths 2017). As Katz (1996: 172) emphasises, it is the researcher who goes in search of knowledge, complete with their research agenda, and when the research is finished, they leave: 'Such moves reflect power no matter what their broader intent is.'

When I returned to my office the following week, the need to analyse data followed me around the corridors. This need had a dark and overbearing presence. I avoided it, skilfully dodging in and out of people's offices to 'catch up': 'How was Bolivia?' was the chorus, and 'fine' was the beat. That was the answer that people wanted, not an emotional outpouring of why I was avoiding my data. Supervisors acknowledged that I would require some time to settle back home and placed little pressure on me to 'be productive'. I had, after all, been 'in the field' constantly, changing it and being changed by it, but the funding deadline loomed and I came to realise that the responsibility to represent a collective narrative lay largely on the fact that I had already conducted my fieldwork, and while the 'field experience does not automatically authorise knowledge, [the fieldwork] allows us to generate analyses and tell specific kinds of stories' (Hyndman 2001: 262). After a few weeks I printed out the transcriptions and translations of my data. 'One step at a time,' I thought. I also purchased some new highlighters online. 'I'll start when they arrive,' I promised myself. I was procrastinating. Fast-forward a few weeks, and with a drafted methods chapter, some theoretical engagements and sections of a literature chapter in draft form, I immersed myself back into the data, starting with what I termed 'safe' data: the government and NGO interviews. In terms of women's life stories, I still found myself avoiding large sections of the audio and transcripts – and in particular the detailed stories of pain, abuse, devastation, brutality and hurt — something that perhaps only those who have conducted similar research on violence will understand. I simply could not bear to return to some parts of the women's lives. I struggled to return to those moments of violence (see Hume 2007) and yet 'what a privilege, to choose to escape reliving these moments when women in Bolivia live them every day' (fieldwork diary).

Doing the 'write' thing

I did not complete the thesis within the three years of funding, but instead sought employment to help me through the final fourth year. I needed the space that an extra year would offer me. This section outlines, firstly, a few of the tools used to address the challenges of moving physically out of the field but remaining emotionally and ethically attached to it, and secondly, one of the many things I would do differently. Upon reflection, I realise that the challenges experienced in leaving the field were very similar to those when I entered: the struggle to adapt, to figure out what I needed to do and what I *should* do, to move to the next stage of research, to develop personally and professionally and to make progress towards completing my thesis. It is only with hindsight that I can see that there were three tools in particular that not only helped me during and after fieldwork but also gave me the confidence to write this chapter. The first of these is a fieldwork diary, the second is drawing and the third is the importance of self-reflective texts from other academics.

Keeping a reflexive fieldwork diary was useful in order to document the negotiation of shifting positions and emotions, and provided a space to bring together thoughts and feelings about my experiences as well as the direction my research was moving in and why (Punch 2012; Chiseri-Strater and Sunstein 1997). Such a task can often seem rather uncomfortable at first, as it encourages the researcher to consider and record, in great depth, thoughts which they may find difficult to confront. Although I had considered keeping my fieldwork observations separate from my reflexive diary notes, in the end I chose to keep them together, with observations written on one side of the notebook (when handwritten) or on one side of the page (when typed) and reflections and feelings on the other. When interview data and fieldnotes are gathered and analysed, it is important to recognise that they are not raw data; they are already encoded with the interpretations and understanding of the researcher (Coffey 1996: 66), and thus I felt it important to capture feelings, understandings and my own experiences as they happened during fieldwork.

Through ethics committees, researchers and participants are often constructed in an under-socialised way. As Blake (2007: 415) explains, the researcher is often thought of as 'simply the vessel into which the subject pours their essence and is [therefore] conceptualized as having no connection with the data produced'. To consider the researcher in such a manner is to ignore the complexity of the social relationships that are continuously negotiated and renegotiated throughout the research. As De Laine (2000) argues, such suggestions underestimate the role that emotions have on the choices a researcher makes and, in turn, the effects these have on interpretations and understandings, as well as the relationships that develop in the field. Engaging with all of this, including emotions, can often draw criticism and, as Finlay (2002: 227) points out, 'researchers are, in effect, damned if they do and damned if they don't'. A fieldwork diary that includes emotions 'in the moment' and not separated from the 'data' explicitly acknowledges the interconnections between the two. Of course, there were times even then that writing did not feel like something I could do, and when I felt I could not write, I drew.

Drawing began in the field as a way to overcome my struggles to articulate and document observations and emotions, and it was something I enjoyed and found peace in. It came naturally. I saw it as doodling, as I do when on the phone, but with more meaning. I did not focus on the quality of the drawings, at least not in the moment, because I did not intend to use them. Aside from one drawing that I featured in a presentation during an event and later at two other conferences, they remain unseen. At the time I did not realise their value, but Causey (2017) advocates drawing as an ethnographic method and technique for translating what researchers 'see' and, in hindsight, I wished that I had analysed and included my drawings in the write-up of my thesis. Even though I did not, drawing helped me to proceed though the liminal space between fieldwork and home, as I continued to draw when I left Bolivia. Drawing and craft became tools for making sense of the ethnographic data when I could not do the 'write' thing. My first engagement with supervisors regarding my findings did not involve a written piece of work. They were perplexed – but enthusiastic – at my initial proposal to do something 'different', and I instead presented them with an interactive paper poster that reflected women's relationships with the law and justice (an image of which can be found in Rogers 2017). Our meeting centred on this document, and I tape-recorded our discussion, of which the transcription provided me with my own explanation of my work, offering me the beginnings of the written words that I had struggled to muster. While at the time I also wondered if I were procrastinating, I now believe that the drawings, the highlighters, the poster and continuing to write in the research diary were strategies of self-care (Rager 2005).

The third tool, reflexive, self-conscious academic literature, grounded me in the knowledge that the feelings I was experiencing during fieldwork and after were normal and valid (in particular Lunn 2014; Punch 2012; Pollard 2009; Hume 2007; Mandel 2003; Ely et al. 1998; Chiseri-Strater and Sunstein 1997). Dunne et al. (2005: 22) suggest that 'reflexivity can partly be understood as the recognition of the social conditions within which the researcher constructs knowledge accounts' allowing us 'to make informed interpretations of what we experience and observe and feel', and while I took comfort in this literature I also returned to literature that quite rightly contributed to the questioning of my presence in the Global South. This included literature on the construction of knowledge, researching Indigenous communities, White privilege and neocolonialism (the most important of which was Tuhiwai Smith 2004). Together, these often left me feeling more conflicted, and, as you can see by this chapter, continue to do so, because, as Griffiths (2017: 5) points out in his relation to his own field of geography, being reflexive means the need for an acute awareness of the 'skeletons of empire and how we [as Western researchers] might embody colonial history in our travels South'. Centrally, in qualitative research that seeks to explore lived experiences, we must ensure that they do not simply become '*data*, grist for the academic mill that become commodities of privilege' (Langellier 1994: 75).

Finally, I believe that the use of participatory action research (Huisman 2008) and the grassroots development of research ideas is central to research conducted in the Global South by outsiders. Funding bodies at all stages must account for this, because much of my discomfort lay not only in my position as an outsider but also as sole researcher on the project. I feel shame that I did not adopt a more participatory approach from the outset, despite funder requirements and expectations at the time. While I put in place ways to mitigate my failure, I remain umcomfortable about my PhD.

Parting thoughts

On a very personal note, this has been a confronting chapter and one that has not been easy to write. Revealing my personal and 'methodological self-consciousness' (Finlay 2002: 210) was an uncomfortable endeavour (Hume and Mulcock 2004). This is not usually the case for me, and in fact I previously organised and presented at an event entitled 'Reflections from the Field' (Rogers 2016). In hindsight though, I was not as open as I would have liked, or as is needed in research like mine, and thus I believe that my openness here, in this chapter, is important for both myself and other researchers. This is not to say that the issues documented here do not exist in other projects, but while there has been a growing advocation of reflexivity and increased transparency in the research process, there are very few spaces created for such engagements either within the broader neoliberal university structures or in research outputs such as reports or articles (Scheyvens and Storey 2003). Issues of an emotional, moral and ethical nature do not have space in the articles written for Research Excellence Frameworks (REF) or for journals seeking to focus on empirical findings – findings that will have been created, pulled through and developed precisely *as a result* of those emotions, failures and quandaries. Given the current nature of academia, with only REF-able articles valued in workload models, there is often little time available to write chapters such as this, and yet they are central to the way we conduct research. Such insights into the *personal* in research are, and will be, silenced, relegated to the backs of drawers and the dusty tops of bookshelves, or perhaps stored only in the innermost thoughts of academics. I have drawn the content of this chapter from all those locations, and yet it still only scrapes the surface, for I am constantly pulled back into the experiences and emotions of fieldwork in random moments. I may have left, but this research experience did not leave me.

There are no conclusions to this chapter because I have never managed to come to any. It is therefore a *pharmakon* piece (Derrida 1981), as my former supervisor Bill Munro would suggest, at once both the poison and the remedy. For although I have engaged with only a *few* of the moral and ethical quandaries I experienced and explored only *some* of the deep-rooted feelings of discomfort that you rarely find in the polished, 'finished' write-ups of my research projects, these revelations – and the ability to write them – are also problematic. They simply reflect and reinforce my privilege, position and guilt. In focusing this chapter on these issues and the situatedness of the research example that is presented, I am drawing attention to the fact that I will not, and should not, escape the title of 'privileged western researcher' (Griffiths 2017: 4). This chapter is not, therefore, seeking to suggest that there are complete solutions to any of these issues, and it is most certainly not suggesting that greater reflexivity, transparency or engagement with emotions will absolve researchers conducting fieldwork abroad. What it does do, though, is encourage more of these open discussions, ones that I have rarely seen in literature from other researchers who have conducted similar work. I confess that I have continued to work in Global South contexts, but now do so with colleagues from within those contexts and with open conversations about power, privilege, colonialism, and the nature of research and funding more broadly. I'm still not comfortable, but I have realised that being uncomfortable is what is needed, and where I should be, in order to reflect on the way I work as a researcher within the opportunities and constraints of academia.

References

Blake, M.K. (2007). 'Formality and friendship: Research ethics review and participatory action research'. *ACME Editorial Collective*. 6(3): 411–421.
Burgess, R.G. (1995). *In the Field: An Introduction to Field Research*. London: Routledge.
Causey, A. (2017). *Drawn to See: Drawing as an Ethnographic Method*. Toronto: University of Toronto Press.
Chiseri-Strater, E. and Sunstein, B.S. (1997). *Fieldworking: Reading and Writing Research*. Upper Saddle River, NJ: Prentice Hall.
Coffey, A. (1996). 'The power of accounts: Authority and authorship in ethnography'. *Qualitative Studies in Education*. 9(1): 61–74.
Daly, K. (1997). 'Re-placing theory in ethnography: A postmodern view'. *Qualitative Inquiry*. 3(3): 343–365.
De Laine, M. (2000). *Fieldwork, Participation and Practice*. London: Sage.
Derrida, J. (1981). 'Plato's pharmacy', in *Dissemination*. Chicago: University of Chicago Press.
Dunne, M., Pryor, J. and Yates, P. (2005). *Becoming a Researcher: A Research Companion for the Social Sciences*. New York: Open University Press.
Ely, M., Anzul, M., Freidman, T., Gerner, D. and Steinmetz, A.M. (1998). *Doing Qualitative Research: Circles within Circles*. London: The Falmer Press.
Finlay, L. (2002). 'Negotiating the swamp: The opportunity and challenge of reflexivity in research practice'. *Qualitative Research*. 2(2): 209–230.
Griffiths, M. (2017). 'From heterogeneous worlds: Western privilege, class and positionality in the South'. *Area*. 49(1): 2–8.
Guillemin, M. and Gillam, L. (2004). 'Ethics, reflexivity, and "ethically important moments" in research'. *Qualitative Inquiry*. 10(2): 261–280.
Hirsch, S.F. (2002). 'Feminist participatory research on legal consciousness', in J. Starr and M. Goodale (eds) *Practicing Ethnography in Law: New Dialogues, Enduring Methods*. pp. 13–33. New York: Palgrave Macmillan.
Huisman, K. (2008). '"Does this mean you're not going to come visit me anymore?": An inquiry into an ethics of reciprocity and positionality in feminist ethnographic research'. *Sociological Inquiry*. 78(3): 372–396.
Hume, L. and Mulcock, J. (2004). *Anthropologists in the Field: Cases of Informant Observation*. New York: Columbia University Press.

Hume, M. (2007). 'Unpicking the threads: Emotion as central to the theory and practice of researching violence'. *Women's Studies International Forum.* 30(2): 147–157.

Humphreys, M. (2005). 'Getting personal: Reflexivity and autoethnographic vignettes'. *Qualitative Inquiry.* 11(6): 840–860.

Hyndman, J. (2001). 'The field as here and now, not there and then'. *The Geographical Review.* 92(2): 262–272.

Katz, C. (1996). 'The expeditions of conjurers: Ethnography, power, and pretence', in D.L. Wolf (ed.), *Feminist Dilemmas in Fieldwork.* pp. 170–184. Boulder, CO: Westview Press.

Langellier, K. (1994). 'Appreciating phenomenology and feminism: Researching quiltmaking and communication'. *Human Studies.* 17: 65–80.

Lassiter, E. (2005). *The Chicago Guide to Collaborative Ethnography.* Chicago: University of Chicago Press.

Lunn, J. (2014). *Fieldwork in the Global South: Ethical Challenges and Dilemmas.* London: Routledge.

Mandel, J.L. (2003). 'Negotiating expectations in the field: Gatekeepers, research fatigue and cultural biases'. *Singapore Journal of Tropical Geography.* 24(2): 198–210.

Pollard, A. (2009). 'Field of screams: Difficulty and ethnographic fieldwork'. *Anthropology Matters.* 11(2): 1–24.

Punch, S. (2012). 'Hidden struggles of fieldwork: Exploring the role and use of field diaries, emotion'. *Space and Society.* 5(2): 86–93.

Rager, K.B. (2005). 'Self-care and the qualitative researcher: When collecting data can break your heart'. *Educational Researcher.* 34(4): 23–27.

Reyes, V. (2020). 'Ethnographic toolkit: Strategic positionality and researchers' visible and invisible tools in field research'. *Ethnography.* 21(2): 220–240.

Rogers, A. (2016). 'Reflections on a year of fieldwork in La Paz, Bolivia'. Reflections from the Field Symposium (University of Stirling, 5 April).

Rogers, A. (2017). 'Claiming the law: An ethnography of Bolivian women's access to justice and legal consciousness' [Thesis]. University of Stirling.

Rogers, A. (2020). '"But the law won't help us": Challenges of mobilizing Law 348 to address violence against women in Bolivia'. *Violence Against Women.* 26(12–13): 1471–1492.

Scheyvens, R. and Storey, D. (2003). *Development Fieldwork: A Practical Guide.* London: Sage.

Stacey, J. (1988). 'Can there be a feminist ethnography?' *Women's Studies International Forum.* 11(1): 21–27.

Sultana, F. (2007). 'Reflexivity, positionality and participatory ethics: Negotiating fieldwork dilemmas in international research'. *ACME Editorial Collective.* 6(3): 374–385.

Tuhiwai Smith, L. (2004). *Decolonizing Methodologies: Research and Indigenous Methodologies.* London: Zed Books.

Index

abandonment 7, 10, 40, 77, 79–80
  analytical task 189
  political 74, 79, 240
  researcher position 114
access 2, 6, 11, 21, 23, 25, 37–40, 63–64, 67, 69, 72, 74, 106, 139–141, 149, 168, 173, 185, 187, 189, 193–194, 220–223, 232, 245, 247
  readers' 3
analysis 1, 9, 24–25, 34, 74–75, 81, 86, 90, 113, 139, 142, 149, 160, 175, 177, 185, 187, 192, 199, 212, 221, 223, 226, 236, 238–240, 243, 247
apprenticeship 20, 23, 86, 90, 152–153, 160–161, 205–213
  informants' 130
autobiography 1, 10, 14
autoethnography 158, 205, 211–212

body 8, 11, 20, 66, 86, 87–88, 89, 91–93, 95, 152–157, 159, 160–162, 207, 209, 211
burnout 76, 177

care 18, 20, 37, 38, 39, 41, 43, 62, 70–71, 89, 108, 169, 175, 238, 247
  self-care 251
career (academic) 2, 22, 69, 79, 116, 157, 159, 162, 221, 239, 243
closure 77, 212, 253
commitment 20, 21, 25, 38, 41, 43, 76, 77, 82, 110, 119, 126–129, 152, 178, 207, 212, 222, 232
community 20, 21, 24, 69, 75–75, 77, 79, 80, 81, 83, 93, 102, 105–107, 110, 119, 130, 136, 141–142, 167, 169, 175–176, 209, 231, 232, 235–236
  of practice 205, 208
consequences 17, 24, 43, 110, 160, 231, 236
covert 23, 120, 142, 209–212
criticism (of informants) 222–223, 231–241

deception 210, 237
desertion 17, 78–79
diary (field) 14, 25, 34, 169, 243, 247, 249–251
digital research 106, 158
  social media 76–77, 81, 83, 239
dilemma 1, 8, 64, 78, 126, 131, 134, 202, 221
disability 7, 66, 70, 107
disengagement 6, 67, 71, 75–76, 79, 82, 113, 192, 200, 202
dramaturgy 232

embodiment 152, 155, 161, 207, 212
emotion 2, 17–18, 20, 33, 37, 39, 64–65, 66, 69–71, 78–80, 82, 103, 114, 118–119, 121–122, 129, 136, 198–202, 211–213, 219, 235, 243–244, 240, 248–250, 252
  anger 80, 16, 131

anxiety 24, 38, 132, 247, 248
boredom 7, 76
desire 90, 92, 95, 96
fear 209, 241, 247
grief 103
guilt 6, 7, 36, 39–40, 65, 69, 75, 79, 82, 209, 211–212
loss 2, 68, 79, 103, 105, 106, 177, 246
pain 2, 10, 19, 38–40, 51, 71, 103, 110, 118, 129, 223, 235, 249
physical 157
sadness 2, 6, 7, 36, 75, 79–80, 82, 247–248
entanglement 17, 18–20, 22–24, 67, 86, 91, 95, 127, 167–168, 170, 173–174, 176–177
ethics 3, 20, 24, 25, 38–39, 41, 43, 86, 131, 135, 141, 149, 169, 202, 243, 244–245, 250
exits (types of)
active 10, 22, 25
'bad'/ imperfect 19, 46, 50, 55–56, 58–59, 200
black hole 9–10, 19, 33, 40, 43, 44
chaotic 35–36, 40–41
forced 2, 11, 25, 89, 113
'good'/perfect 19, 25, 48, 50, 52
labyrinthine 19, 33, 43
planned/managed 2, 4, 11, 14, 19, 25, 36, 37, 52, 70
theorised 9–10

familiarity 114, 118, 143, 181, 184, 195, 201
feminism 75, 78, 91, 97, 246, 247
feminist ethics of care 37, 41, 43
material 87
field
bounded 20–21, 192, 202
extended 118–119, 139–142, 153, 199, 202
fieldnotes 3, 7, 14, 16, 65, 94, 106, 108, 110, 139, 184–187, 222, 226, 237, 246, 248, 250
finishing 7, 67, 69, 75–76, 79, 134–135, 142, 212, 228, 248, 252
friendship 7, 9, 13–14, 20, 22, 24, 39, 64, 66–70, 78–79, 81, 83, 87, 90, 96, 102, 106–108, 111, 114, 131, 156, 159, 161, 172, 211–212, 231, 234–238, 240–241, 244, 246
ethic of 114
faked 195, 201, 210
as method 234–235

gatekeeper 10, 35, 64, 66–67, 222, 224, 226–227, 245
gender 19, 20, 34, 36–37, 42–43, 54, 90–91, 97, 105, 122, 143, 187, 225, 247
femininity 35, 43
masculinity 6, 94, 96, 97, 210
transgender 54, 55
'going native' 189–190, 238

habitus 63, 65, 90, 97, 154–155, 159, 207
embodied 22, 153–155, 212

identity 22, 38, 79, 103, 128, 133, 175, 212, 219
field identity 64, 167, 168, 246–247
occupational 63, 68, 170
professional 37, 40, 67, 75, 76, 77, 169, 174, 176, 226, 227, 239
researcher/practitioner 155, 161, 175–176, 182, 211, 243, 247, 249
impression management 220
indigenous 86, 97, 144, 144, 251
intermission 23, 180–181, 184–186, 190
interviewing 5, 7, 8–9, 34, 37, 63, 74, 76, 93–94, 96, 102, 109–110, 113, 115–116, 126, 130–131, 134, 143–144, 158, 169, 171, 173–174, 175, 193, 222, 234–235, 236–237, 239–240, 244, 247, 249

leaving 2–18, 19–26, 33–35, 40, 46–47, 49–50, 53, 62–63, 67, 70–72, 74–80, 82, 91, 102, 107–109, 111, 113–114, 116, 118, 130, 134, 140–141, 157, 161, 167, 176–178, 180, 180–186, 192–194, 201–202,

212, 228, 234, 239–240,
243–245, 247–249
see also not leaving
love 2, 20, 86, 90–93, 95–96, 247

memory 16, 21, 70, 102–103, 105–106, 109–110, 119, 135, 157, 193, 199, 209, 212
  active reminding 195–196

narrative 34, 37, 39, 68, 109–110, 126, 128–129, 134, 136, 174, 201, 213, 233, 236, 243, 249
  awareness 202
  of leaving 2, 70, 77, 82
  practitioner-researcher 207, 213
  stigmatising 74–75, 81
not leaving 16, 114, 116, 205

participation 142, 189, 195, 211
participatory action research 246, 251
positionality 1, 102, 118, 167, 175, 212, 243, 245, 247
privilege 26, 39, 57, 103–104, 106, 121, 127, 226, 233, 244–245, 249, 251–252
publication 2, 8, 9, 14, 24, 80, 82, 116, 140–141, 184, 223, 239

recursivity 21–22, 140, 167, 173–175, 176–178
reflexivity 82, 102–103, 132, 135, 141, 187, 247, 251–252
relationships 2, 9, 16, 19, 20, 36, 57, 64, 67–70, 74–75, 76, 78–82, 102, 114, 129, 135, 193, 195–196, 198–202, 210, 212, 220–221, 223, 227, 231, 232, 234–236, 238, 241, 244, 250
  disrupted 36–37
  romantic 2, 86–90, 93, 95
  sustained 56–57, 153
representation 5, 24, 63, 70, 77, 81, 83, 102–103, 110, 184–185, 213, 233, 237, 245, 247
returns 2, 5, 8–9, 17, 22, 24, 25–26, 107–108, 114, 118, 131, 156–157, 160, 171–177, 180, 186–187, 192–202, 219, 222, 226–227, 228, 243, 246–249, 251
risk 90, 115, 173, 177, 190, 194–195, 209–210, 225

saturation 8, 12, 16, 75, 79, 221
self 2, 13, 16, 21, 22, 24, 79, 82, 83, 90–94, 96, 120–121, 209–211, 219–221, 226–228
setting 6, 8, 12, 16, 17, 23, 37, 62, 63–64, 75, 78–79, 81, 115, 140, 142–143, 187–188, 193, 198, 208, 220–222, 225–226, 237, 241
  perspicuous 148–149
sexuality 2, 90, 93, 95, 97
  heteronormativity 86, 91, 92, 95, 97
social media 76, 77, 81, 83, 239
stigma 14–15, 80–82, 83

theory 8–10, 16, 86, 133–135, 167, 178, 224, 233

validation 24, 68, 202, 219, 222–224, 228
violence 10, 105, 109–110, 131, 152
  gendered 6, 27–39, 244, 246, 249
  symbolic 225
visibility 48, 91, 92, 181, 183
visual methods 116, 199
  creative 83, 115–116
  drawing 25, 48, 249–251
  photography 34–36, 42, 110, 160
  video 9, 74, 117, 139, 145, 149, 158, 160
voice 21, 37, 108, 117, 119, 126, 129, 134–135, 176, 235, 246

writing 1–4, 7–8, 13, 15–16, 20, 24, 25, 62, 65, 75, 77, 80–82, 86, 92, 96, 117, 121, 129, 134, 136, 171, 176, 178, 187, 189, 222–223, 225, 236, 243–244, 247–248, 250
  vulnerable 106

Milton Keynes UK
Ingram Content Group UK Ltd.
UKHW021640100524
442543UK00005B/36